FREEZE-DRYING FOR THE FOOD INDUSTRIES

FREEZE-DRYING FOR THE FOOD INDUSTRIES

J. McN. DALGLEISH

The J. McNair Dalgleish Resource,
Food Engineering Consultancy,
Warwick, UK

ELSEVIER APPLIED SCIENCE
LONDON and NEW YORK

ELSEVIER SCIENCE PUBLISHERS LTD
Crown House, Linton Road, Barking, Essex IG11 8JU, England

Sole Distributor in the USA and Canada
ELSEVIER SCIENCE PUBLISHING CO., INC.
655 Avenue of the Americas, New York, NY 10010, USA

WITH 18 TABLES AND 62 ILLUSTRATIONS

British Library Cataloguing in Publication Data
Dalgleish, J. McN.
Freezedrying for the food industry.
1. Food. Freeze drying
I. Title
664'.02845

ISBN 1-85166-407-6

Library of Congress Cataloging in Publication Data
Dalgleish, J. McN. (J. McNair)
Freezedrying for the food industry/J. McN. Dalgleish.
p. cm.
Includes bibliographical references.
ISBN 1-85166-407-6
1. Freeze-drying. I. Title.
TP371.6.D35 1990
664'.02845—dc20 89-16945
CIP

Photoset and printed in Northern Ireland by The Universities Press (Belfast) Ltd.

To both our families

Preface

In a multi-basic-disciplined field such as food technology, in order to visualise a process better, one tends to simplify it by imagining a *model* system, in form having likeness or reference to things already known.

Foods then are pictured as 'foodstuffs', and popularly recognised as 'starchy foods', or 'plenty-of-vitamins foods' or 'fresh foods' or other common conceptions, quite adequate for the most part in persuading people to balance their diet and enjoy the prospect of intake and the functions, social or survival, of eating. The complication of enquiring into the means by which we build and maintain our bodies on what has already proved useful and integral to *other* forms of life rarely is considered a subject warranting more than superficial scrutiny from a well-established and secure society.

Retail presentation of preserved foodstuffs is an ancient art, and the general public, provided they are assured of hygienic supply lines, seem singularly un-curious regarding the complexity or otherwise of the highly technical system at their service.

Consequently when, perhaps as a snippet of wrapper news, the public are introduced behind the scenes and asked to accept a phrase describing a preservation process such as 'freeze-drying', they may well be satisfied with the notion that something familiarly edible is actually rather like a honeycomb inside, the combs containing mostly water with a variety of 'nutrients' in suspension, and the whole is first frozen but then dried, after which it is unnecessary to keep it frozen, just dry.

Their model may be amplified with the information that the drying part is the changing of the ice into steam without being melted, and further that atmospheric pressure must be vacuumised in special machines, all of which makes the process slow, elaborate and somewhat expensive, but the resultant quality, when the water is restored perhaps many years later, makes it all worthwhile.

For the layman, such a model is sufficient for everyday visualisation, albeit scant on technical detail. Why should it be contradicted—it has the necessary elements of comfort, assurance of skill, marvel and more than a hint of justified cost.

Alas, for the technically aware, there is much that is glib in this descriptive outline. We know that the chemical composition of any cytoplasm is highly intricate, and that separating out H_2O molecules as ice crystals or in any form is a drying or dehydrating action in itself; that the whereabouts of these molecules within the 'honeycomb' matrix is dependent on the striated fibril layers forming the cell wall boundaries and interconnections; that the nutriments themselves are mobile, but enmeshed in a gel-propertied structure, and their relation to that structure alters depending on the liquidity of their environment. We know moreover that the extraction of the 'moisture' as vapour from the ice phase is dependent more on the partial pressure of water vapour than on the total pressure, and that the phenomenon can take place on a winter's clothesline, in a cold store, or indeed upon the exposed living tissue of an Arctic explorer or a micro-organism, and can render them dead or dormant depending on the conditions.

For the jibe of 'elaborate' one may choose to read 'elegant', for 'slow' substitute 'careful' and for the relative term 'expensive' respond with 'not necessarily'. We are led to explain a much more elegant model, having parameters that fit the constraints of physics and chemistry as we can apply them. The quality then is a measure of the care we have taken, and the cost rests on the engineering skill and the benefits of scale.

Explanation of the construction of such a model for this fascinating foodstuff process, from which to plan optimum performance, is the purpose of this book.

This then is a book dealing with an involved process, of commercial use to a large number of diverse companies in a wide-ranging food industry. It is important, therefore, especially when mathematical treatments of parameters are discussed, to have a concept that is understandable by all parties. Such a vision is not always immediately provable—indeed many of those concepts reported are *dis*provable—but at least the present intention, as no doubt all others, is to conjure a mental picture around which one can argue the practical sense of what is happening in reality.

When a problem has to be programmed to fit the memory or capacity of an available computer (say), then sometimes constraints are applied to the program parameters which cut the problem down to convenient size, without recall. Eventually the workable program may bear insufficient resemblance, in its simplistic form, to the original to enable the operator to use the solution in the real world. Increasingly this is unlikely nowadays, for vast memories are tappable, but (computer) programs are still often built around sub-routines that themselves have crept into common usage, not because of memory size, nor even because of simplicity, but in an endeavour to frame the solution to the 'new' problem in a satisfying familiarity of context.

Thus it has become assumed, almost taken for granted, that one must perform a freeze-drying operation within a hermetically sealed chamber of great strength, from which all gases have been removed to a near vacuum, and that the capital cost of achieving the inherent conditions is high, not to mention the running costs. This need not be!

Dictionary definitions have in their nature to be universal. So it is not surprising to find that Webster explains freeze-drying as 'a process by which heat sensitive materials . . . can be dried by freezing and then converting the frozen solvent to gas in a high vacuum at a low temperature'!

The treatment of blood plasma or antibiotics in medical laboratories is readily pictured by this definition. But what about the classic illustration of the frozen clothes drying on the washing line, without ever thawing? They are still in a vacuum, one might accept, since the whole world is in a near-vacuum, and the temperature profile (compared with the sun) is not all that high. What happens to the vapour? Is it enticed to a colder *surface* to condense again as ice, as many laboratory freeze-driers arrange?

It is not too difficult a coloured concept then to view the Earth perhaps as a gigantic shuttle freeze-drier, or at least a heat pump of reversible type, where the polar cap nearer the sun sublimes to the other, mountain peaks blanch in the wintry wind, and Nature's changing apparel signifies senescence.

Geophysical studies tend to be concentrated at the pole that is Southern, where the landmass, staked out but as yet uncarved, protrudes to point ominously at the sky. Until lately it was mineral content that held the various settler nations' attention, rather than the

(CFC-gathering) hospitality of the heavens (North Pole '89 has also a reported 'ozone gap').†

There it is a delicate balance. The size and amount of minute dust in the outer layers of the atmosphere, where such vapours are super-cooled and invisible, may become enough to permit ice crystals to form and grow before that vapour can reach its ultimate or penultimate destination. Not for it the high peaks of the polar region or even the Himalayas; the crystals are bound for the lower, warmer, higher vapour pressure of the clouds, there to add to the greenhouse effect, urged on by the occasional volcanic eruption or nuclear explosion.

In freeze-drying for the food industry, world-wide, one has to ensure that the ice is not diverted.

J. McN. DALGLEISH

† The Natural Environment Research Council plans (1989) to develop a UK Universities' *Global* Atmospheric Modelling Project, euphemistically shortened to UGAMP, perhaps modestly to signify the Dickens of a large umbrella! In such hands and the data-crunching age, some spin-off on the Spin-Off should prove the adroitness of the move.

Author's Note

When absolute pressure is discussed in this book, it will generally be referred to in mm Hg. This is because a number of other publications, endeavouring to standardise reported observations in Pa or kPa to concede *derived* SI units, have merely converted data in easily visualised *round* numbers, based on scale divisions of 760 mm, into obscure decimal-fractionated numbers that convey little hint of the real accuracy. Thus 0·2 mm Hg has appeared as 26·7 Pa, 20 mm Hg has become 2·7 kPa, and so on. The finesse of quoting a reading on a scale to the nearest mark is lost, and an accuracy is implied which not only was not there but may be misleadingly precise. The ratios 760 mm Hg = 1·01325 bar = 101·325 kPa are given with this proviso in mind (Fig. 1).

A number of non-orthodox but graphic terms have been used in technical contexts. These layman's words are intended deliberately to conjure up more vivid mental pictures for the serious student, perhaps prompting him or her to explore some of the features reported, of which this process has traditionally been embarrassed, and thus to branch into quasi-associated territory previously dismissed as a side-track of assumed less rewarding economical significance.

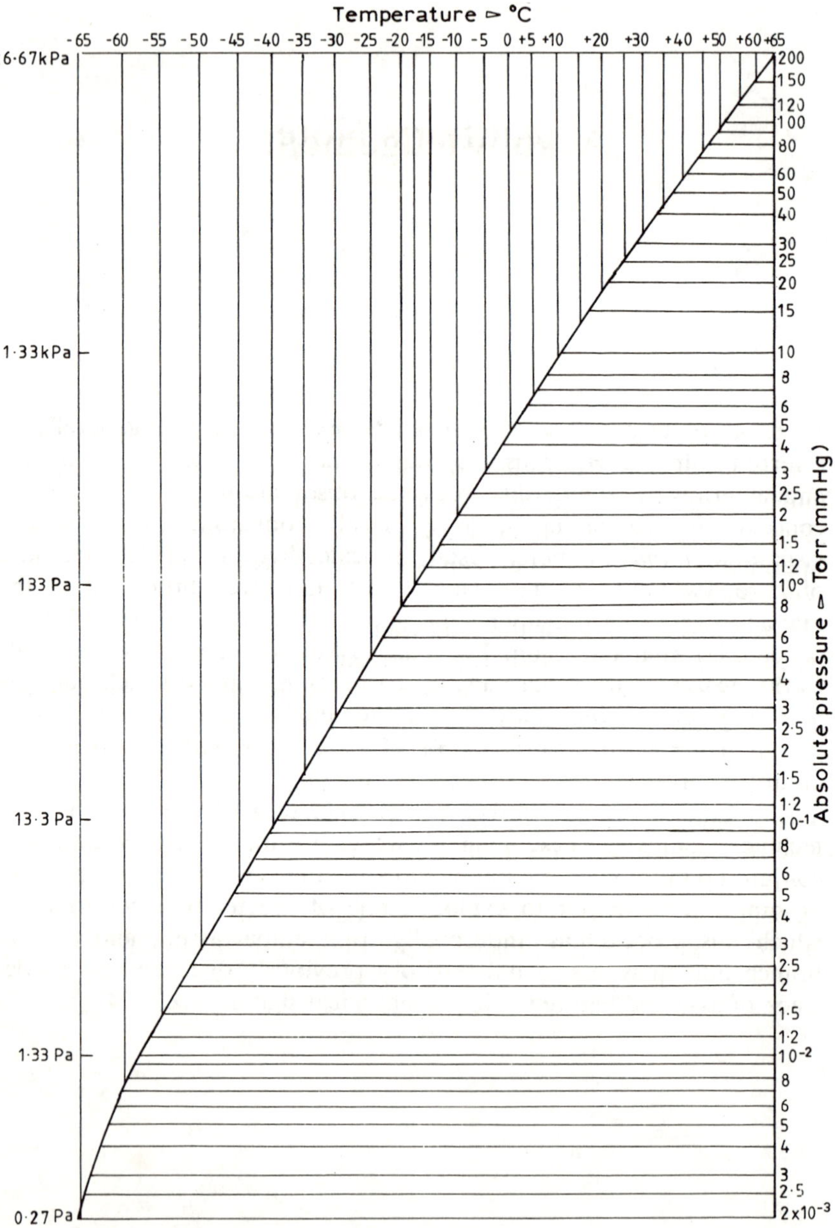

FIG. 1. Pressure/temperature relationship for water and ice.

Acknowledgements

Figs 2, 6, 11, 20, 23, 25, 29, 49, 50, 53(a) are reproduced by kind permission of the Controller of Her Majesty's Stationery Office. Source—Ministry of Agriculture, Fisheries & Food.

Figs 3, 4, 10, 18, 24, 34, 35, 47, 48 relate to the author's work at the Experimental Factory, Aberdeen and the Ministry's help is hereby acknowledged.

Thanks are due also for the following contributions:

Fig. 31 from A/S Atlas, Ballerup, Denmark

Figs 27, 28 courtesy of WGG Cuddon Ltd, Blenheim, New Zealand

Fig. 7 & Table 4, Irish Sugar plc, Carlow, Eire

Figs 30, 36, 37, Table 14, Leybold-Heraeus GmbH, Hanau, FRG

Fig. 9 Mars Ltd, Melton Mowbray, Leics., UK

Table 5 Haldane Foods Ltd, Barrow upon Soar, UK

Table 6 Science & Life (1985) courtesy "Intercosmos Council", Moscow, USSR.

Fig. 55 Table 7 (1970) NASA & Natick Labs., USA.

Fig. 60 Crown Copyright Official Army Photograph 88-04-21-33.

Contents

PART IV **Futures**

Part I

HISTORICAL BACKGROUND

(*EVOLUTION OF COMMERCIAL FOOD FREEZE-DRYING*)

Chapter 1

The Role of the Aberdeen Experimental Factory

The second half of this century has seen considerable emphasis placed on foods that are quick and easy to prepare, whether the feeding pattern was in famine relief, or in combat zones, or in highly urbanised societies. Tastiness, wholesomeness, variety of diet, reasonable cost, with no waste and with ready availability, outline the consumer demand to which industry has responded by increasing preparation *within* the factory of top quality ingredients and materials, and stressing the subsequent convenience at the point of consumption—the pickling and canning stories all over again, perhaps, but both public and food industry were shaking off the fetters of Second World War rationing. A new look was sought. As with all forms of food presentation, assurance had to be inherent of freedom from bacterial spoilage and off-putting enzymatic changes, while flavour and mouth-feel, original or acquired, had to be acceptable and enticing. Preservation had to start at the farm gate, distribution chains within and without Great Britain were primitive, motorways in their infancy, grocery shops still small and multi-outlet.

Freeze-drying came near to being dismissed out of hand as inordinately expensive by the food industry, not the general public. They in fact were demonstrably eager, in post-austerity, to try all things new, from instant soups to TV dinners. To the entrepreneur processor, unaided capital outlay appeared too high for this process, which for a start had a long production time. Commercial quality tight-roping for a just-adequate packaging form was unknown, and some considerable extra marketing effort, compared with (say) frozen foods, was deemed likely. Many socio-economic factors were at work. Government, commerce and public had different attitudes.

Nevertheless, freeze-drying of foodstuffs, in general much cheaper than for pharmaceutical products, and therefore less able to bear a high value-added cost, was thoroughly researched and developed. The advantages made it very attractive, technically. The whole process is conducted at relatively low temperatures, there is no shrinkage as with hot-air drying, so the light and porous pieces retain their shape and dimensions. The bulk transport of distribution and storage is filled before it is laden, unless compression is resorted to.

Easily reconstituted, such pieces *look* desirable and retain 'bite', while flavour and aroma come back remarkably well—highly persuasive attributes! In the event, it was soluble solids of infused beverages which brought freeze-drying back 'out of the cold' as it were, rather than the recognisable slab steaks initially striven for. Even so, the sheer cost of maintaining the cold chain as a national and international network is beginning to remind commerce of that early metaphorical fork in the road.

HISTORICAL BACKGROUND

The Role of the Aberdeen Experimental Factory

In 1948 the Commonwealth Advisory Committee on Defence Science (CACDS) considered, among such glamorous topics as atomic warfare, electronic controls, imperial manpower, etc., the subject of food supplies, and realised that, with the end of the war, both research and production of food products of special military and civil defence application had ceased.

The original members of the Commonwealth Committee on Defence Science–FOOD were (from Ministry of Food HQ) Dr N. C. Wright, Chief Scientific Adviser, Dr H. R. Barnell of his department, Prof. A. W. Scott, Engineering Adviser, with G. N. Lawrence, Dr K. Bryce Jones, S. W. F. Hanson and J. Parker, (from the War Office) Major Radcliffe, Miss Lowden, Mrs Seeking, along with Mr Hayes of Army Operational Research Group, and (from overseas) Brigadier N. N. Chopra for India, Lt-Col. T. M. Niaz and Lt.-Col. N. R. Adhami for Pakistan, J. H. Hulse for Canada, Dr V. Armstrong for New Zealand, Dr R. C. Hutchinson for Australia and Dr F. W. Quass for South Africa.

The Indian delegate proposed that research and development on dehydration should be the priority, and this was adopted, as was the proposition that any developments should be capable of proof at factory scale level.

Reflecting on World War II, it was clear that the forces feeding systems, and particularly the commandeered food factories, provided in emergency, had been hurriedly put together, depending notably on converted textile driers and hot-air drying generally. Consultations were held on long-term co-operation between Commonwealth countries to forestall any future problem of the kind.

The major responsibility for revitalisation was placed on the shoulders of Dr Norman Wright, then Scientific Adviser to the Ministry of Food. Accordingly he was additionally appointed Scientific Consultant to the War Office on food matters, and he and Major-General H. M. Whitty (as Inspector-General of the Royal Army Service Corps) visited American and Canadian service units associated with research, development and procurement of foods.

It was accepted that dehydration was the significant food defence process; it had made enormous contribution, but its limitation problems were just becoming clear when the war ended. [Research in this area had been undertaken by the Food Investigation Board (director Dr Franklin Kidd), with establishments at the Low Temperature Research Station, Cambridge (animal products), the Ditton Laboratory, East Malling (fruit and vegetables), and Torry Research Station (fish). Procurement had been by Dehydration Division (director G. N. Lawrence) of the Ministry of Food, advised by 'Jimmy' van den Berg of Unilever.]

The recommendation of CACDS led to the setting up of a Commonwealth sub-committee, still comprising representatives of India, Pakistan, Canada, New Zealand, Australia, South Africa and the Rhodesias. The whole effort at this stage was of military complexion, and the application of the food-related sciences to military intelligence purposes made for logistically strange bedfellows.

Among the reports of BIOS (British Intelligence Objectives Sub-Committee) and other US and Commonwealth sources had appeared news of a novel type of vacuum-drying process invented and developed in Denmark but set up for production of dried fish in Norway.

Although, as stated above, dehydration in the United Kingdom had effectively ceased, there still remained, within the Ministry of Food, a few members of Dehydration Division, whose director (Mr George N. Lawrence) was sent to Copenhagen while Dr H. R. Barnell (Dr Norman Wright's deputy) and Prof. A. W. Scott (Engineering

Consultant to Dehydration Division) went to Aalesund in Norway to see the prototype, which was the only existing plant.

This equipment had been set up as a factory by the Germans, during the occupation, to provide dehydrated fish for their armed services but it had never operated, the Norwegians finding it impossible to oblige. It did work, however, but was not releasable. The Danes said they had learned enough from building it to make a better one, and consequently a Preskfisk plant, as it was known, was ordered for Commonwealth study.

In the UK an organisation was set up in the Chief Scientific Adviser's division of the Ministry of Food, with Dr Maurice Ingram (ex LTRS) in charge. S. W. F. Hanson (ex TRS) was asked to spearhead a projected research establishment in Aberdeen, Scotland, and a United States/Canada liaison post was set up in the British Joint Services Mission in Washington (first occupied by Price-Davies).

The closest establishment in format to the system the committee had in mind was Torry Research Station in Aberdeen, Scotland, which had full-scale arrangements for studying the harvesting of deep-water fish. It also had spare land nearby. Money at that stage was not a problem, but over the next few years the admirable co-operation links became less distinct.

India's independence included setting up her own R & D centre in Madras. Pakistan had its own problems around that time. South Africa became a republic in 1960, and withdrew her application to rejoin the Commonwealth as such. The New Zealand Scientific Office devoted much attention to food research while Val Armstrong was liaison officer, and its engineering industry, anxious to go it alone, was able to supply its needs in that direction anyway. Canada also had the nucleus of a dehydration industry. The residual co-operative centres were CSIRO in Australia and Ministry of Food (swathed in the combined Ministry of Agriculture, Fisheries & Food) in the UK. All this 'entropy' was in the future, and unforeseeable.

With the hard-won experience during the war of convincing commercial firms (whose food processing factories had been commandeered for the war effort) to institute improvements in technique as soon as they were evolved in the government laboratories, it was realised that developments had to be seen to be able to proceed to full-scale prototype demonstration, if indeed *years* were not to be lost in transfer to commercial production in peace-time. From this remarkable prescience the research establishment was to have, as an integral

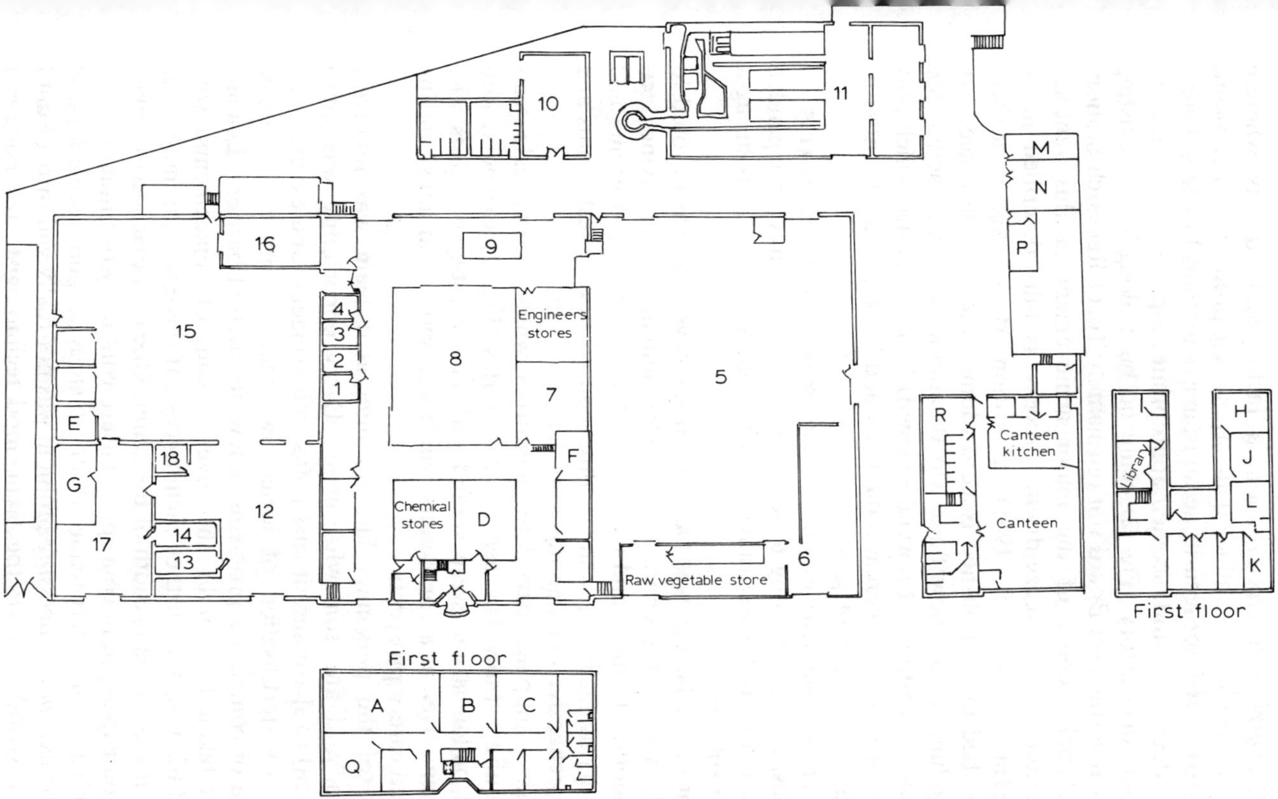

FIG. 2. General Plan of MoF/MAFF Experimental Factory, Aberdeen, Scotland, 1950–1960. Laboratories: A, research and analytical control (meat and fish); B, research and analytical control (fruit and vegetables); C, bacteriological laboratory; D, bacteriological preparation room; E, packaging laboratory; F, control laboratory (vegetables); G, physics laboratory; H, J, antioxidants fundamental research; K, vegetable research laboratory; L, fundamental research laboratory; M, N, fruit and carotenoids research; P, cobalt-60 source; Q, design office. Factory: 1, storage room (desert conditions); 2, storage room (jungle); 3, storage room (tropical); 4, storage room (temperate); 5, West Bay, fruit and vegetable air-drying; 6, root vegetable preparation; 7, vegetable pilot plant; 8, engineering workshop; 9, drawing office; 10, wood and plastics workshop; 11, boiler house; 12, fish/meat preparation; 13, chill room; 14, cold room; 15, vacuum/freeze-drying; 16, homogenising room; 17, packaging; 18, carcass conditioning room.

part, an *experimental factory,* i.e. a factory-scale facility wherein experiments could be conducted incorporating production-size plant. This was to be sited, for security reasons, in a potentially low-target area, and Aberdeen, Scotland, seemed appropriate, especially as a custom-built laboratory/factory (Fig. 2) could be built alongside an existing Fisheries Investigation Board station, namely Torry Research Station.

The official account of the major achievement of this venture, Acceleration of the Freeze-drying Process, is fully described in an HMSO (Hanson, S. W. F., 1961) publication. Much other work there was described in unpublished Record Memoranda. For instance, one memorandum relates how flour was de-aerated and compacted into blocks about one-third of normal (bagged) bulk, a feature which may still interest storers of cereal mountains and perhaps put the nose of numerous rodents out of joint.

Another development was of a steam scalder or blancher (Fig. 3) whose design of condensing towers (Fig. 24) led to study of expanded metal as a heat and mass transfer device. The story of this is brought out in Chapter 6.

Layout of the factory was based on three bays; one was devoted to vegetable and fruit preparation (coupled initially to air-drying facilities), another to meat and fish preparation (coupled to freezing and vacuum-drying facilities), and the third housed storage test rooms and the comprehensive engineering workshop.

An agent supplied part-time operatives when factory-scale tests were required. Individual aspect laboratories (Fig. 2) extended into adjacent buildings and were staffed with enthusiastic specialists who found themselves in a community which had short-term aims offering telescoped career prospects.

Fusing of the personnel into an integrated team was relatively straightforward and somewhat novel. Every opportunity was given (and taken) to share and discuss plans, achievements and experiences, so that cross-fertilisation of ideas, awareness of other disciplines, reference of compound problems, knew no internal barriers. Leadership was taken in turn by the several heads of departments, who provided the thrust to further each success. If there was an indigenous problem it was in the breadth of the remit. One eventual success, alas, proved too expansive in concept to ensure continuity of funding.

The collation of expert teams jointly to study all ponderable aspects of a particular *food processing* problem was not really widespread until well after World War II; the term 'food technologist' had not been

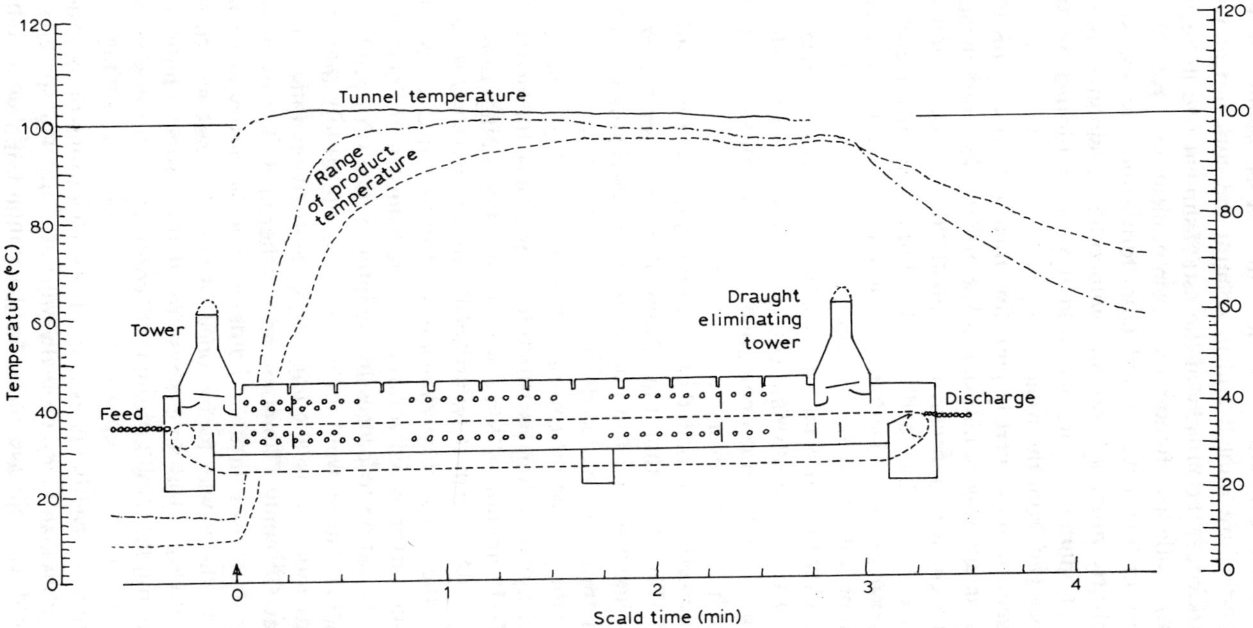

FIG. 3. Prototype steam scalder (for enzyme inactivation). A slatted conveyor took trays pre-loaded with vegetable pieces to be dried, through a long chamber, where superheated (i.e. dry) steam played upon them from jets above and below. Sulphite spray dosing was also possible. (To prevent steam spillage, and cold air ingress, which would have curtailed effective length, special end 'towers' were designed.)

invented, let alone evaluated. Within the Experimental Factory, scientific methods were pooled and in general the common aim, the *size* of the workforce, the *novelty* of the establishment and its relative remoteness, all made for freedom of communication. Over the 10 years or so of its existence, much cross-fertilisation of ideas was achieved simply by rubbing shoulders with other disciplines. A new multi-disciplined, perhaps more rounded, type of intermediate level investigator emerged from the mould.

Heads of departments were eager for this to happen, previous compartmentalisation was minimised in the belief that breakthroughs had then more chance of occurring. Equally there was pressure to produce positive results by deadlines; lifetime study prospects were neither encouraged nor indeed for long envisaged. Consequently members of staff with such background, attitude and training were eagerly sought after by industrial concerns when it was announced that the structure was to be dismantled. The Ministry also felt that broadcast sowing of the human components was an end in itself. The publicised emphasis, however, was squarely placed on the fast freeze-drying aspect of the dehydration remit. Indeed it was officially announced (MAFF, 1961) that 'the advent of the Accelerated Freeze-drying Process had *concluded* the work of the Experimental Factory, which would therefore be closed down'.

There were other factors of course. The annual budget covering the activities of nearly (including part-time and servicing) 100 people had risen above £1·25 million—'What was a Ministry doing running a *factory* anyway?' (Although it was officially recognised as a Research Establishment, the name *Experimental* Factory—coined to contrast with the wartime factories dehydrating for the Ministry of Food which frowned on modification to processing as interference with production schedules—stuck.) There was never any real suggestion that it be handed over to industry as a going concern—a co-operative on the lines of that at Chipping Campden or Leatherhead, for example—simply because there was virtually no dehydration industry as such in the UK and, if there was to be one, MAFF did not see further foundation work (Fig. 4) as an expense for *that* ministry. The market was *not* ready-made. No government contract of any size was forthcoming.

The armed forces—captive consumers for such rations on occasion, but now keen on civilian fare—were dwindling in size (Fig. 5), and the Ministry of *Food* was still associated in the public mind with civilian

FIG. 4. Label of introductory can of AFD foodstuffs.

rationing. The leisure, picnic and snack-food markets were not then in vogue. Thus the potential was prospective and had still to be worked on.

The Aberdeen exploration had tackled the preservation by dehydration of a great many individual foodstuffs for combination in the dry state, and had created a number of complex recipes with culinary fervour, to the extent perhaps that the impression which was given out declared the AFD process to be a food processing panacea (*Aberdeen Press & Journal*, 1960). This had not been the original intention, nor

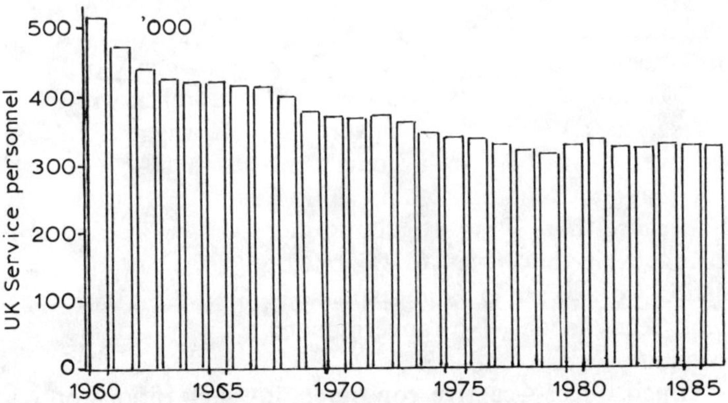

FIG. 5. Strength of UK Regular Forces, *including* personnel with service station ready access to civilian fare (source: Central Statistical Office Monthly Digest).

the spirit of the invention, which had been specifically designed and directed at the rapid freeze-drying of large *steaks*, i.e. thick slices of meat or fish (Fig. 6) having major plane areas convenient for two-sided processing. Certainly it was a bonus to find that loose particulate material could be treated without much trouble as large beds or mats at beneficial speed, but it was never the intention, for instance, to pass the method off as ideally suited to thin soups and sauces frozen into slab form, at least without adaptation. So a number of disappointments ensued when (independently) pilot plants were provided to product development centres of persuaded food processor firms.

The request for large steaks of meat had come from Service sources, who sought to have a ready-made network of suppliers fully equipped, which could be called upon if and when the need arose. The existing food processing industry, on the other hand, was reluctant to indulge unaided in new equipment, beyond pilot plant scale sufficient to test-market a limited range. It was having to concern itself with catering for the needs of wives going out to work as well as their

FIG. 6. Fish slabbed for fast freeze-drying.

spouses—convenience in feeding certainly, but tending towards package soups, whisked desserts, TV meals, frozen foods as such, say—and the attractive accent seemed to be on possible proliferation to encourage the idea of transferring some cost to a storage chain involving the social appeal of ownership of domestic refrigerators. Consequently they saw their role as tending to go as far as factory methods of quick freezing, or discrete particle drying, and encourage what consumer markets there might be along these frontiers.

Guidance of households on food matters is now a significant feature in daily papers as well as weekly or monthly magazines. Then it was sufficiently rare and vituperant as to form a major force shaping public opinion.

The sudden death of Dr J. M. Hay, FRS (Edin.), engineer of repute, marked an unexpected change in direction of the Greyhope Road Station. Instead of heading towards concentration on food process engineering, of which accelerated freeze-drying was an initial step, and although that facet in no way was diminished while the planned budget forecasts were tenable, biochemical research was felt to be at least as necessary.

This took some of the uniqueness away. Such work was in competition with other centres, strong on benchwork; notably the Research Establishments which were funded in part at least by trade subscription and which each had a governing body composed of representatives from commercial enterprises. These not unnaturally thought that public moneys should be used to help them directly and, if not, be expended on research lines agreed communally by them. The name 'experimental *factory*' (Fig. 3) was a relic from the wartime days when the Ministry commissioned air-drying food factories to preserve foodstuffs. Experiments therein clashed with tight production schedules and full-scale experimental runs on novel equipment were not possible in such circumstances. Scientific Advisers Division longed to be in the position of demonstrating unhampered that its findings were capable of quality production on factory size equipment and under authentic conditions. However, the powerful days of conscripting industry had gone. Why should a government department be doing things this way? This was the era which had seen the conversion of the Cambridge Low Temperature Research Station, the DSIR, the Campden laboratories and the Leatherhead complex. Where should research become development and where should development stop?—when the government was absolutely sure that private industry would take it up *and follow through* without further support, or when the government knew enough detail itself to institute an emergency programme again should the need ever arise?

Feelings and pressures of lobbies were strong. There were a number of centres which sought identity and backers. The experimental 'station' at Aberdeen, next door to Torry Research Station, the Marine Laboratory, and within hailing distance of the Rowett (animal husbandry) Research Institute and the Macaulay Institute for Soil Research, was caught in a political compromise. It could stay open amid severe economic cuts and pressures if it could find an independent subscriber. This it did in the shape of Armour, the Chicago meat-packing company, who were prepared to have factory-scale plant designed, built and tested in Aberdeen so that they could acquire comprehensive know-how.

Food rationing in Britain ended in 1954. The merger of the Ministry of Agriculture and Fisheries with the Ministry of Food, to form the Ministry of Agriculture, Fisheries and Food (MAFF), took place in 1955. The food side of the composite ministry, which did not become a cabinet post until January 1957, was no longer the dominant sector, with the continuing desire to expand farming. Our overseas foodstuff suppliers could invest in dehydration if they so chose. So the announcement (*Scottish Daily Mail,* 17 July 1956) that 'Scots scientists invent the meal in a matchbox—synthetic food, as tasty and even more nourishing than the real thing . . . has been achieved under conditions of top secrecy by government scientists . . . at the Ministry of *Agriculture*'s food research station in Aberdeen,' —did not go down well. The article, by a *science* correspondent, went on to say that the only thing to be added to the chemicals(!) was water, and that research was continuing on producing completely dry chemicals that would produce 'food' when mixed, *without* water!

The next day, to an even wider circulation, Cassandra of the *Daily Mail* wrote: 'Unlike the honest egg, the conspiring cheese, or the dewy lettuce, this extraction of slag-heap muck stays put and can be swallowed for months afterwards with all the calories and the whatnots remaining scientifically intact'.

Clearly the knives were out, to aid the forks! The Ministry of Defence, whose eye on the war in Korea had been the basis of a slightly earlier account in the *Manchester Guardian* of the work, was unabashed. Two Secretaries of State for War were moved in succession to MAFF, but by then the wind-up was preordained and all that remained was to implement the machinery of sublime dissolution.

Some indications of contingency planning were gradually announced. In an article in the local *Press & Journal* (March 1960) entitled 'Your menu for the moon' (thus foreshadowing the July 1969 results of the work done by Natick Laboratories in the USA), S. W. F. Hanson (officer in charge at closure) was reported as saying 'The important thing really is not that British firms alone should go in for dehydration, but that countries from which we import should use the process', and on another occasion: 'Taking a long view for example, we may question whether it is necessary to convey New Zealand lamb carcasses halfway round the world in special refrigerated ships carrying bones and the water contained in the meat, when one might reduce the carcass to consumer portions, freeze-dry the water out of these, and convey the wholly edible lightweight product without special ships or warehouses'.

The minister decided that the processes had been 'developed to a point where commercial exploitation should be undertaken by private enterprise' and that the establishment be closed on 31 March 1961! The premises were sold to Unilever, the ministry's agents, who 'propose to continue to undertake research into problems of food'. Some of the team stayed on there, but it was not long before the study of freeze-drying and kindred foodstuff problems were being investigated elsewhere.

In June 1961, two members of the British government (one of them the first of the two Secretaries of State for War earlier referred to) visited Commonwealth capitals to discuss the possibility of Britain joining the Common Market. The initiative passed amicably to 'Vickery's mob' at the Commonwealth Scientific and Industrial Research Organisation in Australia. Sadly, when the time came, the EEC denied freeze-dried New Zealand lamb (meat only) special entry facilities, above those for whole carcasses, to the Common Market, despite the strenuous pleas of the Meat Industry Research Institute of New Zealand, and other interested bodies. Lamb now arrives chilled. Nevertheless, or perhaps because of the widespread controversy, the process of freeze-drying for foodstuffs as such had strode on to a world stage.

Chapter 2

The Spread of Activity

The disparate gap between food process technology and food product technology in the eyes of the business world was nowhere more apparent than during the 1960s surge of interest in freeze-drying.

Most food processors on Service lists had had to revise their production plans considerably during the previous decade and compete for the public's favours in the novel realm of convenience. Their favoured operation on a cost-plus basis for supplies to government contracts which entailed substantial outlays in capital expenditure was no more, and the climate was not ripe for an abundance of these. The engineering concerns that had indulged in considerable development programmes, in expectation, were faced with a wary food processing industry bereft of investment capital sponsorship as well as large dependable orders. They had to guess how such firms might calculate their costs, and persuade from limited knowledge how the venture might be set off. Pilot plants for product development purposes were in demand, and samples of orthodox dried products were produced in plenty by equipment manufacturers and offered to potential processors together with suppositional cost schedules as best could be furnished.

Some processors, on the other hand, while eager to have their product development teams trained in (or at least familiarised with) the new techniques, were unwilling to indicate what the style of their possible new range of products might embrace, or perhaps simply reluctant to channel or limit their innovatory ideas at too early a stage. Thus the momentum of the initial enthusiastic reception was prolonged in a cautionary arcane campaign rather than be left out, which proved expensive and testing.

Indeed one company, already well versed in design and manufacture of a wide diversity of driers, agricultural and chemical, in air and under vacuum, but from the liquid phase, built an experimental freeze-drying pilot plant with full ancillaries into a mobile demonstration and teaching van, for use on existing food factory premises by temporarily released personnel. Not one tangible order directly resulted, although concurrently they had sizeable success with a set of batch driers for medium strength coffee concentrate.

It was a 'you first' situation rather than 'me too'. The simple message that accelerated freeze-drying brought, that a process previously thought of in terms of 48 hours or so, and now capable of drying 15 mm thicknesses in less than an 8-hour shift, *must* therefore be cheap enough for foodstuffs, was in danger, just because costings on an undisclosed market scale and certified by valid commercial evidence were extremely hard to come by.

Choice was also complicated by the very eagerness to be competitively diverse, of the engineering concerns themselves. The 'expanded metal insert' process, which had been patented in only the UK and the USA, was thought of as very much dependent on the earlier plate closure scheme. Indeed, Atlas of Denmark used the AFD acronym to signify *Atlas* Freeze-Drying as well as *Accelerated* Freeze-Drying. The stress initially laid on the ideal suitability of the system for steaks or really large slabs of regular thickness was played down in the claim to be a universal process, or rather a process eminently suitable for beds of sundry shapes of cut foodstuffs. The one process that enthusiastically kept to the slab theme was the Canadian 'spiked plate' scheme but, even when hollow, these spikes required to be sufficiently close together such that removal of product seemed to present an added mechanical complication on a factory scale with some concern about cleaning, which slowed acceptance.

The question of vapour disposal represented a major decision, and a variety of car-loading systems were being advocated almost before emerging from the novel testing stage. Industry was keen but wary, and there were few unbiased and knowledgeable enough to guide it in its decisions. The plunge was taken by the Irish government via General Costello, a man of prompt and broad decision, to build, on the recommended lines, a (Vickers) four-cabinet sequence-batching factory-scale installation, for a wide band of home-grown seasonal produce, and with communal steam ejector extraction. This effectively smoothed the parabolic if not asymptotic style entrainment, making steadier duty for supply steam. The Mallow plant performed continuously until old age claimed it in 1986.

In Italy, government and other assistance via the Casa de Mezzogiorno enabled the Leybold prototype tunnel freeze-drier at Salerno (in actuality twin shuttle-chambers) to come on stream, and German Federal moneys prompted the farmers (milk) co-operative at Dahlenberg with the first true tunnel plant. The Unilever subsidiary Batchelors Foods at their Ashford, Kent, factory also installed a Vickers AFD set of cabinets operated on borehole water and steam ejectors, consolidating the pay-back period for capital by introduction of a risotto range, much appreciated by the military.

Despite sustained campaigns worldwide, most of the other installations were, besides pilot plants, single-cabinet units for some time. In the USA no European plants were sold, despite considerable activity, beyond of course, the two Atlas-built plants designed by the Aberdeen, Scotland, personnel for Armour the Chicago meatpackers.

For the post-Aberdeen decade there was considerable marketing activity amongst the few engineering companies who ventured wholeheartedly into this field and could stand the pace, despite the range of specialised expertise it called for.

Atlas pilot plants have been delivered to the following spheres of influence: Angola, South Africa; Argentina (2), Chile, Cuba, Ecuador, Peru, Colombia, Venezuela; Bulgaria, Poland; Denmark (12), Finland, West Germany (2), Iceland, Italy (3), Portugal, Switzerland (2); India (2), Indonesia, South Korea, Malaysia, Philippines (2), Singapore, Taiwan (6).

The Leybold group delivered somewhat larger pilot plants to Belgium, Britain (2), France (2), Germany (3), Netherlands, Norway, Hungary and Sweden. They already had a number of cabinet plants operating with finned dishes, extending to one at their Cologne headquarters which became an entry lock for their new tunnel system.

Vickers concentrated on Eire and the Commonwealth; Mrs Kruschev and family tasted their AFD products at the British Trade Fair (1961) in Moscow. In that year they built the Irish Sugar (Mallow) 4-cabinet plant, and two cabinets for Batchelors Foods (Unilever) in Kent, all with steam-driven venturi extraction systems.

Atlas in 1961 built six production AFD cabinets for Maggi (Nestlé) in Germany for soups, two for Afico (Nestlé) in Holland, one for Liebigs Extract of Meat Company in Sligo (later to become a Unilever plant) and one for an undisclosed buyer in New Zealand, all with steam-driven extraction.

The original Atlas 'vacuum contact dehydration' (VCD) cabinet, converted to AFD before specialised freeze-drying cabinets of that process had been specified, was sold from Aberdeen to the tea division of the Lyons group at Guildford, complete with its German extraction gear.

Thereafter it is difficult chronologically to trace all the production plants, of which there were a considerable number. Some exchanged hands, some were modified after receipt, and some were copied or rather elements were assembled to do the job on an *ad hoc* basis, partly to keep aspects of certain food processors' pre-market-launch activity confidential, and in other cases as an endeavour to produce on a cut-price basis.† Abroad there was no longer a dearth of interest in freeze-drying foodstuffs, as Tables 1 and 3 indicate.

Vickers, Atlas and Leybold approached freeze-drying of foods from different backgrounds, but remained competitors. Leybold's strength lay in non-condensables extraction systems, with ultra-high-vacuum and clinical freeze-drying their previous forte. Atlas, up to then, were principally a 'freezing for the fishing industry' company.

Vickers Armstrong (Engineers) were the innovative part of a marine/aeronautical/weapons conglomerate, dependent on Whitehall, Commonwealth, Allied and Overseas government contracts, and were looking for diverse pursuits during a peace lull. Their venture, for instance, into tractors had not been profitable. They were as yet only testing the new nuclear propulsion machinery in simulated submarine conditions at Dounreay in Scotland, coincident with the building of Dreadnought, designed for long self-sufficient voyaging. All their establishments, however, were at capacity except the old Spitfire, later Scimitar, aircraft works at South Marston, near Swindon. The management there took on several diversification projects besides AFD, including postal segregating and sorting equipment (Evans, H., 1978), which brought them into direct competition with Elliott Automation, Leybold's partner in Britain, and the American market. They tended with AFD to follow the lead given to Atlas by the original MAFF constraint which required improved vacuum drying (later freeze-drying) to stem from the double-contact system.

Vickers South Marston Engineers applied themselves successfully to the main problem which accelerated freeze-drying posed—that of

† A considerable proportion of the large number of pilot freeze-driers was retained by front-name food processors with no intention of owning full-scale units but who wished to guide their tied contract suppliers on best processing procedures for their secretly designed special products.

dealing with very high commencing and, at the end of the cycle, very low quantities of extracted vapour—by designing, in conjunction with Hick Hargreaves, a four-cabinet sequence-batching installation which evened out the (extraction) duty, and *proving* it (Fig. 7) for the (government backed) Irish Sugar Company at Mallow. This relatively large order, early on, ensured further input of venture capital.

They did a similar exercise for Unilever (Batchelors Foods, Ashford, Kent) but, apart from a prototype continuous operation ('tunnel' plant) for Brooke Bond on tea/coffee extract (Brooke Bond/Liebig later consolidated into Unilever), they returned to AFD single cabinets on automatic (cam-contour) control. They had come up against a phenomenon quite strange to their normal way of business—they had to develop a market amongst their customers' customers, and without government aid (Lefever, 1960).

Heinz UK, who had been co-operating with Catelli of Rossi & Catelli, Palma, Italy, on hot-break (pectin retained) tomato liquor

FIG. 7. Quick reconstituting snack products such as made by Erin Foods, division of the Irish Sugar Company.

evaporation to paste (and had a Vickers pilot freeze-drying plant), supported R & C design of a novel freeze-drier with overhead ice-condensers and pumping equipment by the Italian branch of Edwards High Vacuum. (Fifteen years later Heinz opened a can of freeze-dried vegetables, packed under N_2 at this plant, and pronounced it excellent.) When Heinz Marketing created Heinz-Erin in conjunction with the Irish Sugar Company, they withdrew from the Italian freeze-drying venture at Torrechiara, although their on-site process engineers remained there.

Meanwhile Leybold had established *its* prototype 'tunnel' freeze-drier on the outskirts of Salerno. Launched by Casa de Mezzogiorno, the office for the relief and development of southern Italy, and skippered by advisers from Batelle University, SACAF (Società Azionaira Conservatzione Alimente Freschi) de Fuorni was a 'trimaran' with weakness in its other two hulls—wine from local grapes, vacuum-evaporated concentrates from local citrus fruits—and had run temporarily out of goodwill. (SACAF in due course became SACAF(F), with a new refrigeration facility, enabling it to re-group its existing resources and enter the dried coffee concentrate business.) Nevertheless this installation (like ISC for Vickers) proved adequately to Leybold that their designed scheme of continuous operation was practicable and stable, if not fast. Despite its complicated financial and administrative environment, and reluctance to resort to unexotic, humble members of the onion family which offered profitability in the growing area of dried soups, the SACAF double-lock worked as an engineering concept, and Leybold were able to further this elsewhere.

Vickers, as a group close to bureaucracy, had a habit of forming new subsidiary companies to follow up any project that seemed at all likely to succeed, leaving the original venturing subsidiaries to struggle on tighter reins, or perish. It was more than blinkering R & D. The Polaris submarine programme of the 1960s, while involving several facets of the group, did not seem to connect or ring a dinner bell with the food preservation and Service rations concept being advocated as a separate division at South Marston. Criticism in the House of Commons of excessive overspending on the submarine fleet may not have helped. (Even today it would appear that the Turbulent class of hunter-killer nuclear powered submarines, otherwise splendidly designed to rove under the ice and from ocean to ocean, stock up with 'fresh' (frozen) food supplies to keep them going for a mere 6 weeks (TV announcement on Turbulent class submarines, 17 July 1987). At least they have their own sea/fresh water conversion system, so it *is* possible to extend their effective voyage by use of compact freeze-dried rations.) Whatever (if any) the reasoning, all these vessels were fitted out to depend on cold rooms and refrigerated food stores (witness the interesting human story of the *Swiftsure* cook inadvertently trapped under the weight of it all).†

Also, in connection with expansions at Barrow, the Vickers board dealt with expenditure on 'conventional' canteen facilities in place of those 'found impossible to obtain from a frozen food manufacturer as originally planned'. Clearly the marketing of Vickers pilot AFD products was not even then penetrating, let alone pervading, their sister organisation!

The nice balance between ambitious pioneering in technologies in which little or no expertise existed within the parent organisation, and commercial prudence eager to axe loss leaders, was bothering several sections of the group, and some questioned whether indeed it should not resume the role of *aircraft* manufacturer.

Concorde was not so far off over the horizon. AFD seemed very small beer.

Vickers had been approached by Seffinga, of 's-Hertogenbosch, Holland, with a design of a semi-continuous freeze-drier, but on rationalisation had decided 'Freeze-drying is not strictly within the pattern of our business as a whole and is not directly complementary to any of our other interests' (Board minutes of 15 September 1964, Vickers Armstrong (Engineers), South Marston). They had failed to clinch an earlier arrangement with Ameliorair for agency rights in France and areas of French influence, but were collaborating (through their 40% stake in BAC) with various supersonic aircraft. So, when the Schneider group, an immense organisation with many interests, took over Delattre-Levivier in 1965, later fitting it into Creusot-Loire (including Socaltra-Levivier), it was not difficult for a food processing venture of Socaltra to connect (Oct./Dec. 1966) with South Marston.

For a consideration, which included adopting or adapting manufacturing agreements with Buckau R. Wolf in Germany and Japan Vacuum Engineering Company in Japan but in money terms

† Leading Cook Griffin survived after spending 80 min buried under 200 frozen chickens, 50 lamb joints, 30 huge chunks of beef, all at −40°C in its deep-freeze store, when the nuclear submarine unexpectedly tilted through 30° (Anon., 1985).

amounted only to some two-thirds of Vickers 'losses' so far, and on a 5-year sales projection of about five times what Vickers had achieved (Vickers' own private projection was much less), the freeze-drying hardware and know-how changed hands.

At that time the French government was offering, to French companies who engaged on projects that would process home dairy products, 50% of research costs, plus a gift of 1/3 the capital of the processing company, with a further 1/3 at agricultural rates, leaving only 1/3 to be put up by the venture. If AFD was to have a further chance to develop in design, surely this was it!

Rights to the British and French patents covering AFD, assigned by MAFF to NRDC in March 1960, were not however renewed by fees in 1967 and cover therefore lapsed. The Minister of Agriculture, Fisheries & Food for 4 years from July 1960 became Ambassador to France in 1968, with plenty on his mind, but Vickers' possible option to re-enter the freeze-drying field was not taken up.

The Rank Hovis McDougal group meanwhile went quietly about their own system of freeze-drying egg and similar products, using Atlas freezing equipment only. The resulting powder had exceptionally good blending quality, an important factor in their baking business. In a book of cases and comments on consumer product innovations A. Robertson (1970) describes the food freeze-drying efforts of the Irish Sugar Company on their Vickers plant; based on information which an MA student had gleaned mainly from BFMIRA, Torry Research Station, Unilever and Erin Foods (later Heinz-Erin), it makes a number of points, the main one being that Erin Foods faced competition from Unilever, Corn Products and Nestlé. They concentrated on ingredient use for freeze-dried items, maintaining their other lines, whereas ISC produced dried solid foods direct for a retail market undergoing relentless persuasion towards similar but frozen foods before the frozen food chain in the UK was even fully linked (Fig. 8).

Erin Foods were advised to take their quality control—American style—all the way back to the farms whose produce they were promoting by co-operative processing, under the aegis of the Irish government. A broader line of products was recommended and sought. Their marketing drive was eventually linked (1967–8) with Heinz, who boosted sales to 180% within a year.

In 1968–70 Erin Foods opened another freeze-drying factory, at Midleton, Co. Cork, and equipped it with two cabinets, this time by

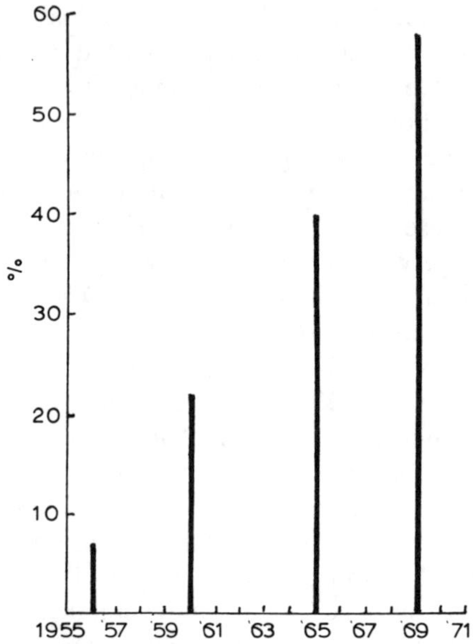

FIG. 8. Domestic refrigerator ownership per UK household (British Refrigeration Association, pers. comm.).

Stokes, having ribbed trays suspended between the fixed heating shelves on a mono-rail carrier. Extraction was, and still is, by ice-condenser and vacuum pump. The Mallow plant continued in diversified service until 1986. Latterly it was operated with its movable plates fully open, with relatively deep beds of particulates. Thus freeze-drying as such, a production cost within complex raw material purchasing and complex export marketing, survived the throes of marrying the farming produce not only to the market but to the Market. It was Vickers who succumbed. Some feed-back warned of hiccups, but did not deter (cold key comfort). One group that was especially thorough with its cost fact-finding was the Mars organisation, who still regularly look for novel processes as a means of furthering their already well established lines of foods for domestic animals. At that time they were seeking to can recognisable whole pieces of fish, and were experimenting with frozen raw material.

The Scandinavian vacuum dehydration process on which the MAFF Experimental Research Establishment centred its initial below-atmospheric-pressure investigations, was called Preskfisk and did just that—pressed fish. However, fish muscle fibres (Fig. 15) run longitudinally, and overlapping fillets stifled drying, so transverse sections of cod (as in salmon steaks), with the belly flaps tucked in, were a practical sequel. Samples of freeze-dried cod steaks were stacked, neatly without much movement, in suitable round tin canisters and distributed to potentially interested processors by the Ministry. Amongst these was Petfoods, who ran cat acceptability tests in March/April 1959. Favourable reports came from owners who expressed a preference.

Some owners did not cook the samples, merely reconstituting them in cold water to the intended 5 ounces (142 g), a matter of a moment or two, yet to the cat the temperature or rawness seemed to make no difference. There was felt opportunity to promote the product on both quality and convenience grounds. Single steak packs were planned in flexible envelopes of the style then appearing for frozen foods. These pouches were to accommodate a 5/8 in (16 mm) thick piece not more than 5 × 4 in (125 × 100 mm) overall.

Cod, which had been used at the prime fishing port, Aberdeen, by MAFF, was mostly larger in cross-section than ling, allied to the codfish, long in body and dried as 'stockfish' (1958 annual UK catch 5000 tons). The central body portion configuring to such sizes was found on average to be 53% of the purchased weight, but Petfoods could utilise the rest on other lines. There was some seasonal fluctuation in availability and therefore in price, resulting in bulk buying and frozen storage.

Atlas agents in Aberdeen, who formed Kestrel Foods in 1954 (later Aerated Foods), had set up a pilot plant with 15 trays 1·67 m × 0·75 m, i.e. about one-third the size of the Ministry's plant, to exploit the process commercially. Petfoods obtained sufficient samples for a two-town test market between November 1959 and May 1960.

Results showed that 12% of cat-owning households bought the product Flick (Fig. 9) regularly (about 2·4 packets per week) while existing popular brands seemed unaffected. Several price levels were tried and a ceiling was not found, although the company reckoned that 1 shilling (£0·05) was probably then a reasonable retail price limit on which to base projections. Translating this to a national volume, 6 months after introduction, Flick gross sales value would be running at a rate of just under £1 million per annum. A substantial market existed for the product.

Unfortunately there then began a period of uncertainty. A decision

(a) (b)

FIG. 9. Front and rear of prototype flexible pouch for freeze-dried product (0·001 Polythene, 15 lb GIP paper, 0·009 mm foil).

to close the ministry's 'Experimental Factory', to some extent brought on by the name, had been held over on the intervention of Armour, the Chicago meat-packer, who wished to take up AFD in a number of ways. They actually funded the factory (research establishment by then turned process development base) to remain open in return for a business advantage and advanced specification for an AFD cabinet. In fact they ordered three such cabinets, the first one to be installed there in Aberdeen for a wide-ranging series of production trials, the results of which necessarily were required to be made publicly available. The corollary was that Atlas could not supply anyone else with such cabinets until all the factors in the Armour agreement had been fulfilled.

The ministry's agents on-site, who supplied the workforce for production runs, were advising (Fig. 10) a process-only dehydration cost based on 24 cabinets producing 1000 tons of dehydrated foodstuff per annum on a 24 hour, 6 day week. Depreciation was spread over 10 years. The parent company's internal practice at that time was to write off capital cost of similar plant over half an expected life of 11 years.

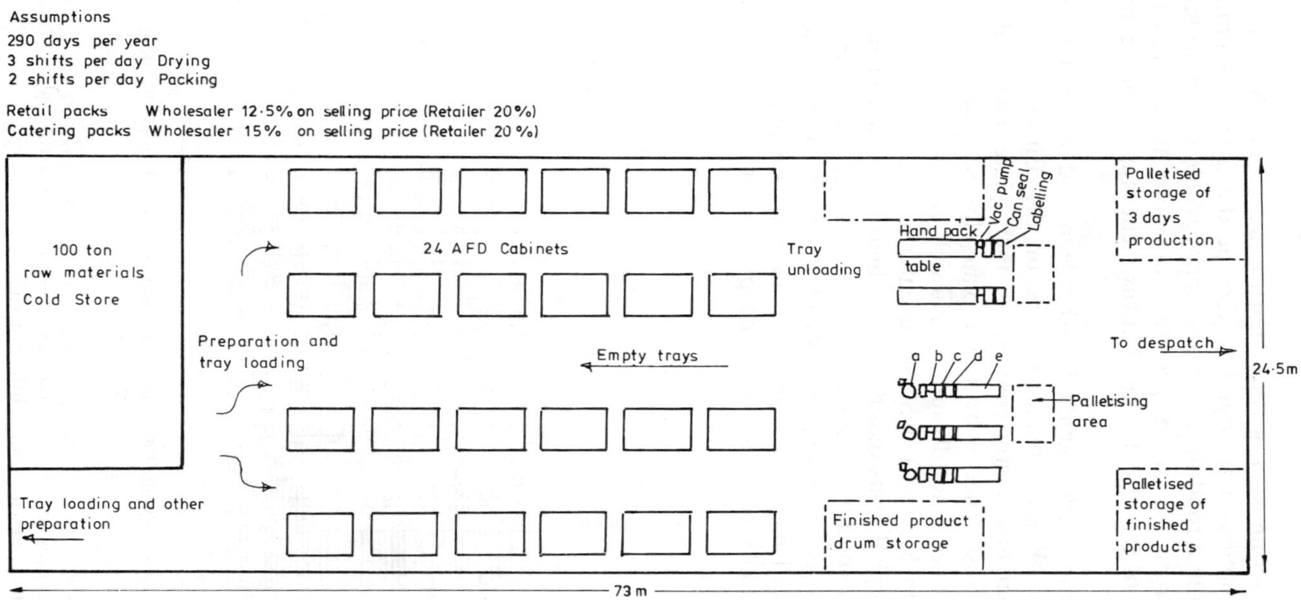

FIG. 10. The agents' plan (surely a strong hint that viability depended on capacity). a, Auto weigh; b, vac pump; c, pouch seal; d, labelling; e, cartoning.

Mars' practice, not unusual, was to amortise in 5 years. This tended to tone down the optimism surrounding genuine throughput assurances, although normally a food processing company would expect to tune a plant after purchase and in private win more than guaranteed production out of it. Clearly initial investment was assuming prime importance.

Meanwhile, the Atlas agents in Aberdeen had entered into an agreement with Winget, manufacturers of the Birdseye 'froster'. Atlas's VCD patents mostly centred on their plate-spacing devices (Fig. 11). The froster had a hydraulic plate-closing system (Fig. 12) which could produce a not dissimilar sandwich effect.

Vickers (South Marsten), who were entering the lists around this time, forecast capital costs based on 4 cabinets somewhat higher than other tentative quotations from pharmaceutical freeze-drier manufacturers. Leybold were pointing out that their system, although slower, could be rendered more versatile,† but in any case did not involve plate closure, and overall 'should be cheaper in first cost'.

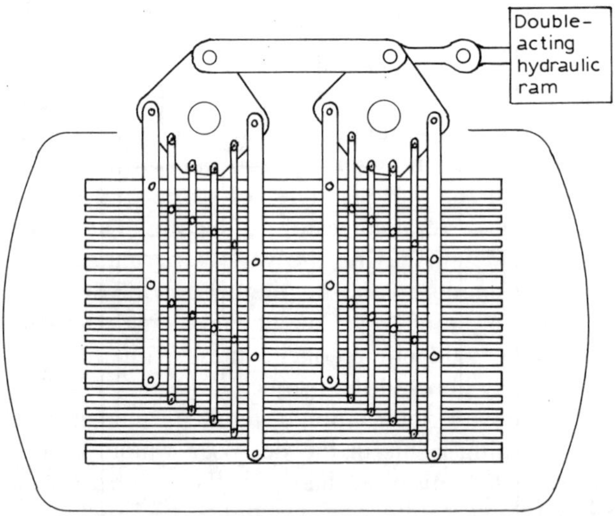

FIG. 11. Original plate-spacing mechanism in double-contact vacuum dehydration system.

† River Wreake water at Petfoods headquarters was too high in temperature for vacuum augmenter steam cooling. The vapour extraction had to be by other means or the operation on a different site.

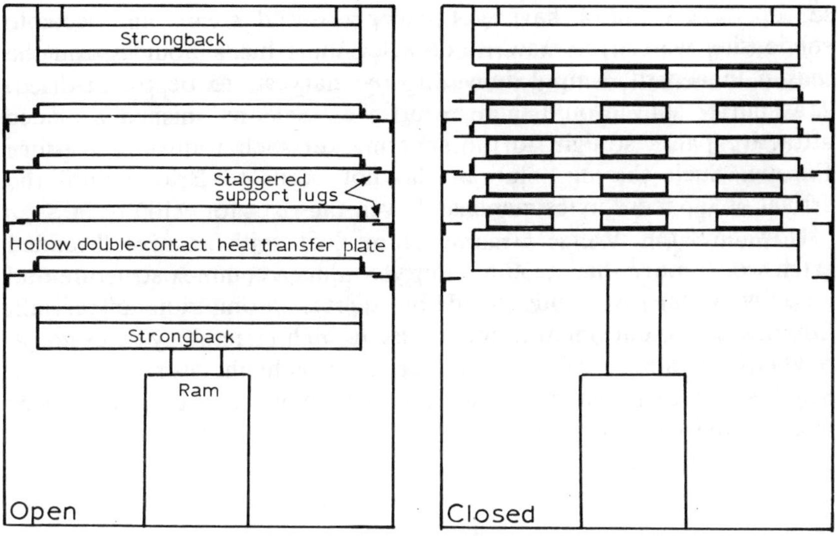

FIG. 12. Plate movement system which relies on product as spacer. Note that the weight borne is cumulative.

In all this atmosphere of healthy competition it seemed provident for Petfoods to concentrate on other aspects. They refined the package, checked out more retail price limits, considered and commissioned trial batches of (fish-finger type) rectangular blocks which were neater to handle on the various styles of plant, and had advantages and disadvantages at several stages of production. They computed full factory costings in detail, having the various permutations, and even considered improvements to the heating system of their own design. By mid-1963, when capital costs of all makes had come quite close together, they still could not reconcile these with their own projections, and reluctantly aborted the project.

Some time later an international feud, termed the Cod War, developed over fishing rights in the deep sea waters around Iceland. Diplomacy did not feature perhaps to the extent that it should. (British fishermen eventually gave up and drifted on to cosier oil-rig jobs, still in the North Sea, while the purpose-built freezer-trawler vessels were disposed of in the southern hemisphere.) No one taking part in the matter seemed prepared to conceive (let alone concede) that Iceland itself, quite closely linked with Denmark's interests, should be the place where a modern freeze-drying facility should be set up. There, to a haven of cheaply raised steam and ice-cold condensing water, the trawlers on safer short hauls would spend the season in accord, simply delivering the harvest, to be freeze-dried. Transporter ships would then export the steaks as finished goods in attractive packs straight to the supermarket shelf routes of Western Europe. Such sharing solutions did not occur to Britain until the arrival of Japanese investment, and even then not for AFD.

Nowadays the diverse UK sea-fishing fleet, with an overall landed catch worth more than £250 million/year, has a complex structure and capacity which is manipulated by quota, prohibition of certain fisheries, and inducement to modernise to high technology by grant. A predictive computer model (Shalliker, 1987) is being developed by the Sea Fish Industry Authority to test and compare fleet-management policy scenarios.

Chapter 3

Emergence of Techniques

It was with considerable difficulty that commercial freeze-drying of foodstuffs wrenched design away from tradition, whereby pharmaceutical and medical specimens were treated with clinical asepticism on the laboratory bench or multi-ampoule-phial-flask up-scaling. Indeed, for a time, penicillin was dried in the enormous 2-m diameter spherical glass vessels in which it was grown.

On the very sound basis that the product would undoubtedly command whatever added-value price public demand put upon it, no serious attempt was made to hasten, understand or even control the process, partly also because the comparableness of working at different temperature levels below that which might separate the substance from its glass lamina was not an objective. Besides, the difference in quality between vacuum-drying (i.e. from the liquid phase) and true freeze-drying was just too great to be contemplated.

Means of telling during actual freeze-drying when a substance had reached the stage of being fully and adequately processed was not strenuously sought, an allowance of up to an extra 100% in cabinet time being not unusual. Expenditure was more justifiably spent on polished stainless steel and hygienic sterilisable accoutrements. The legend was not queried (in fact the mystique was promulgated) that only very expensive or precious commodities could be considered. Freeze-drying was just too good an upmarket process for the likes of foodstuffs; as then practised, this was true.

Most of these vital materials were in the form of slurries which could be frozen on to clean glass by a swirling action to give a thin uniform layer. The thermal conductivity of ice ($0{\cdot}53 \times 10^{-2}$ cal cm^{-1} s^{-1} °C^{-1}) is about three times that of glass, so that room air warmth around an

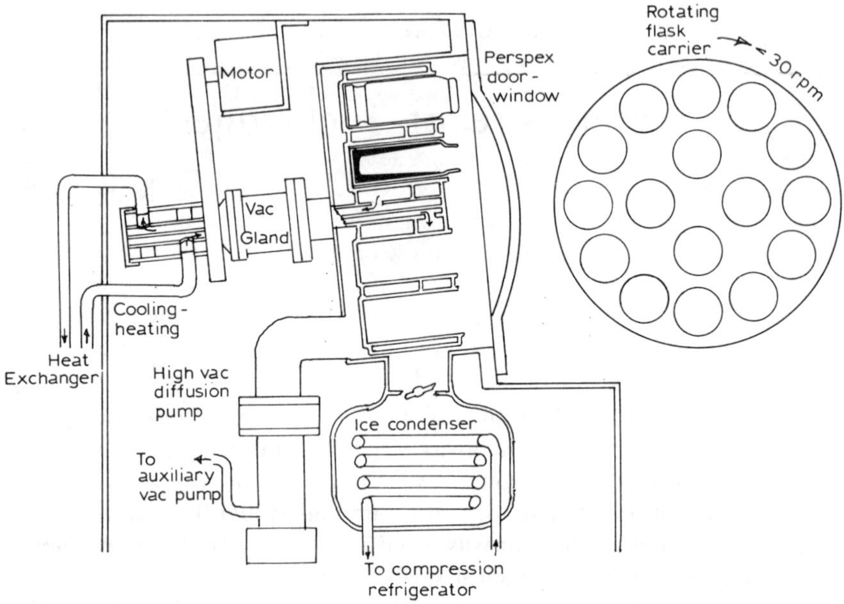

FIG. 13. Common features of early (sterile) freeze-drier: (a) freezing, drying, sealing, all *in situ*; (b) thin layers, light loads, gentle treatment; (c) freezing to −25/−30°C or lower; (d) high vacuum; (e) warming only just above 0°C; (f) several days processing.

ampoule attached to an extraction system is even, gentle and safe. The ampoule can afterwards be sealed off before detachment.

Such habitual thinking led to rotary machines for mother's milk (Fig. 13) and, in chamber drying, to flat, shallow trays on double-purpose hollow shelves. In the first instance coolant was passed through a meander of pipework in the shelf until the material to be sublimed was frozen securely to its tray. Then, under vacuum, the refrigeration machine was 'reversed' in circuit so that the shelf was heated. The alternate switch at the end of drying defrosted the so-called ice-condenser.

Typically chambers were built into walls and had doors at each end so that loading and unloading were effected in different sterile rooms. Inert gas of hospital purity was regularly employed. These stringent isolating procedures are to this day a far cry from the gathering and sun-drying of herbs, the age-old sources of medical remedies and concoctions, although an understandable progression of refinement, as more and more was learnt about what actually was required to be preserved, and how best to do it.

Parallel history has been slower in the cases of fish, fruit, meat and vegetables, simply because the end product had to be of acceptable recognisable size.

Drying of whole gutted fish and substantial slivers of meat at the mercy of the elements has persisted world-wide despite the inevitable limitation of shrinkage to hard tough texture. Tree and vine fruits, allowed some degree of fermentation of natural sugars, have actually grown in popularity. Forced-draught hot-air drying has brought assurance of a certain regularity of quality to many products which can thereby attain a safe water activity in harmony with their prospective environment. With each step of technical skill and know-how came further necessity for attention to detail and perhaps round-the-clock manpower, adding up to a noticeable cost. Air-drying traditionally centred on the harvest surplus.

It is therefore no great surprise that other forms of preservation—canning, freezing, pickling—which shortened the routes from grower to factory, and made easier or planned the network of distribution, enjoined co-operative expansion and the public's involvement. Venture into vacuum-drying or indeed freeze-drying, when it occurred, was restricted largely to bulk drying, near abattoirs, of animal glands, later to be pulverised into pharmaceutical powders. Before hormone science was precisely identified, sweetbreads were more than a delicacy.

Much later came interest in perfecting ingredients for instant vegetable soups, dried gravies, fruit yoghurts, all of which were sought and specified as tiny though discernible fragments and yet were processed in the main with as few cuts during their wet state as possible, i.e. as larger pieces. Retention of culinary quality competed with dimensions of space and therefore time. The large flat 'shelf' had to cope with easily spilt, randomly spread, discrete sections of odd shape. Equipment to suit had to be devised.

Two major approaches were made which became basic to the next stages of development. One was the use of *high*-sided trays (Fig. 37) with equally high longitudinal divisions, say 7·5 cm, about 10 cm apart, and 0·75 m square, of sturdy aluminium, but still quite heavy.

This allowed awkward materials such as shredded cabbage literally to be poured into the tray, patted down and levelled off to a volume fill which gave a remarkably consistent weight, and loaded into the chamber by one operator. On removal, the robust tray could be emptied quickly by overturn and rapping. The spaces were wide enough to be cleaned easily and be seen to be clean, and the 'fins' or partition plates conducted heat via the base to and from the shelf, thus penetrating the mass of load. (Contrast this with the limitations of cross-draught air-drying trays, with no sides and low loading per square metre.) Vegetable pieces which had previously been scalded were easier to handle and pack down, and the loads were heavier.

When mechanised manipulation evolved, the trays became smaller, three to the half-metre square, and were constructed of precision (aluminium) extrusions. The spaces between the 'fins' were narrowed to about 25 mm centres, and automatic filling, emptying and cleaning systems were devised. Some suppliers settled for car-loading by hand of complete batches of trays, and these trays were intermediate both in overall dimensions (and therefore weight) and in rib division. In every variation, the deep tray, which had to be accommodated in wider shelf spacing, i.e. fewer shelves for a given size of chamber but *not necessarily* less load per chamber, put more burden of heat and load on each unit of shelf area. Yet the nominal distance from heated surface to centre of ice was no longer half the depth of the tray, but half the width of the rib spacings.

Coincident at this stage were arguments about the effect of radiant heat versus or in accord with contact-with-shelf heat. Since there was an interleaving of trays with shelves, and some systems advocated tray-carrying cars or trolleys which slid into place without the trays actually resting on the shelves, the heat from both top and bottom surfaces of the shelves had to go somewhere. The maximum temperature then contemplated was 100°C, more commonly 80°C, at which levels the wavelength of the most copiously emitted radiant energy is in the near infrared around 80 000 Å. A radiation of about $3/4\ \mathrm{kW/m^2}$ could be expected. However, this would reduce rapidly during the course of the run. If, for instance, the radiating surface were at 140°C, the significant wavelength would be 70 000 Å. The resultant *radiant* energy flow between a surface at 140°C and one at −20°C would be in the order of $1{\cdot}44\ \mathrm{kW/m^2}$, falling to zero when and if both approached a common 'safe' terminal value of (say) 50°C.

The other approach (Hanson, 1961) stemmed from advances in vacuum drying (i.e. in the liquid phase) of production quantities of fish fillets. These were (a) to make the flat trays so large that edge spillage was relatively a lessened hazard, (b) to concentrate on a product whose uneven shape and thickness, although sizeable, could be dove-tailed by astute operatives into a fairly even mat, and (c) to suspend the heating shelves on a double-cantilever mechanism such that they could be brought closer together hydraulically in sets, thus providing double-*contact,* above and below.

Each flat-tray load was prepared by hand on a large work-table covered with a protective aluminium cover, slid—all 8 ft × 4 ft of it—by two people into the appropriate slot in a mobile carrier, and in due course three sets of seven tray-loads were trundled and winched into the cabinet. Considerable physical pressure could be applied during the run, smoothing out the inevitable shrinkage. The resulting 'sheets' of fish were sawn into pieces about 14-in square, placed one on top of another in an 800-ton press, to create a solid block of dried fish. Thus maximum reduction in volume as well as weight was achieved for storage, although reconstitution was embarrassingly slow.

In both of these approaches the trays were initially made as big, or the tray-loads as heavy, and the shelves in as high stacks, as could comfortably be managed by manual workers. The load per run was stretched, but the duration of the process was not of itself a prime target. Presumably 48 and 24 hours were just as preferable as sub-multiples thereof. In any case, intensive packing of the foodstuff, while increasing the areas of contact with the heat-feeding surfaces, tended to stifle the escaping vapour. In the case of liquid-phase drying, a cycling routine of plate closure and release, giving discernible improvement, was not difficult to determine (steam ejectors were used to remove vapour), but with the finned dishes (some designs of which could be suspended above or laid on to their heat sources) the initial foodstuff contact conditions prevailed.

While quite high differences in level of temperature could be tolerated when ready flow of vapour from cut-cell surfaces of product meant a rapid drain of (latent) heat, this advantage soon diminished owing to the combined extension of heat and vapour paths within the material, plus the requirement not to inflict prolonged high-temperature detrimental effects on already dried surfaces.

In the early 1950s it was not customary to insert a device such as a thermocouple into a product in a production plant since they were either of laboratory apparatus delicacy or, when sturdy enough to withstand expected rough usage, far too cumbersome and blunt to give anything other than a spot check, the typicality of which could not be confirmed without many more sensors having the same disadvantages.

So the main reliance was placed on repetition of experimental uncurtailed runs adjusted for overall time by end-condition inspection and custom, coupled with observation of the rates of fall-off of cabinet total absolute pressure.

It was consideration of how to achieve refinement in this latter feature that prompted investigators at the Experimental Factory in Aberdeen to tune the extraction gear on the Preskfisk vacuum cabinet to a degree of performance that brought the absolute pressures attainable below the nominal 4·58 mm Hg which marked the threshold of freeze-drying possibility.

An interesting corollary of scaling up or more truly comparing notes between pilot plant and full scale, without employing a myriad of thermocouples, was adoption of the interrupted run. Laboriously but painstakingly many series of pilot scale routines were halted at progressive points along the cycle and not only was a general weight-loss curve determined but also cross-sections of product were cut open to reveal the recession of the drying front within the piece. In spite of the complexity of overlapping layers of fish fillet, it was possible even in that case to construct a 'drying plan' for a production run with reasonable forecast of the outcome, but scrutiny of heat and vapour co-dependence in detail became obligatory.

The heating plates in the Preskfisk cabinet were constructed from two 3-mm thick mild steel plates held apart by rectangular steel bars which formed a labyrinth for the heating fluid. The welding of this hollow structure made for great rigidity but also slight undulations. To minimise external rust the plates were regularly treated with (vegetable) oil which, at the initial temperatures obtaining, did not survive for long. Both the tray and lid, of 3-mm aluminium, were subject to fairly hefty handling simply because of their size, and did not remain perfectly flat. Added to all this, the fish fillets, even skinned, varied in thickness from about 25 mm pectorally unevenly to less than 9 mm near the tail. The muscle strata as filleted did not make for short internal vapour paths even had they not been overlapped towards an average tray-load depth of 25 mm, as measured by eye.

Despite these practical factory-scale approximations, almost to be expected, the plant produced an adequate product of the nature and quality intended. In further attempts at streamlining performance of the particular equipment, which had been provided as an available study starting point, great reliance was therefore placed on the incipient compression of all these layers to even out the supply of heat.

Unfortunately this seemed to be the opposite timeous intention to what the designers had had in mind. Early severe closing of the plates, before there was an ample dried-out scatter of portions amenable to resilient compression, tended to raise the product temperature locally above that corresponding to the absolute pressure in the 'adjacent' (*sic*) chamber. However, because of the trapping, isolation, or simply poor connection, and therefore limited external vapour paths, the product moisture did not vaporise until the physical plate pressure was alleviated, whereupon the heating system was unable to maintain the required increased rate, and the intensification of the process lapsed to a lesser activity. So either 'hunting' took place or a compromise nearness had to be sought which permitted periodic advancement of both plate closure and process to be made. The use of expanded aluminium inserts (Figs 6 and 35) as a means of achieving a practical inexpensive solution to this problem has been told at length elsewhere, but the smooth acceleration of what had now become a freeze-drying process had still to be examined.

A case was made out for an intermediate size plant, small enough for economic trials, large enough to provide adequate samples, on which all the perceived and even some of the suspected variables, if not actually measurable, could at least be observed more closely.

Both budget cost and concentration of effort expediency governed the policy decision to continue with steam-driven evacuation equipment rather than refrigeration plus backing pump which was being utilised, if not intensively studied, by kindred research centres. Hence the eventual associated extraction equipment posed no limitation on the attainable rate of drying in this intermediate plant, being (as was thought) considerably oversized. Attention was devoted to optimising what could be done with a set of parallel heating plates, as wide but only half as long as those in the parent vacuum-drying double-contact cabinet. Details of EVE and subsequent cabinets are given in the official HMSO handbook.

Besides the obvious advantages of being able to draw weight-loss curves coincident with plate pressure, plate spacing, heating water temperature (which could show a drop along the plate), cabinet absolute pressure, and record continuously product temperature at predetermined vertical depths within the same piece by means of a gang of thermocouples, and of being able to see, through large windows and good lighting, the product at *all* stages, it was hoped to learn, by virtually full-scale yet inexpensive interrupted drying, why some phenomena occurred—such as unequal dry layers on upper and lower surfaces, nearly dry material being cool or warm to the touch, and liquid phase adjustment to nearly uniform moisture content.

The realisation that expanded metal inserts made possible the freeze-drying in record time of thicknesses previously barely contemplated as commercially feasible led to concentration on cross-cuts of larger fish fillets or even bone-in cod steaks, and beef, pork or mutton muscle, all pre-frozen and bandsawn to an even layer dimension. Nor was foodstuff unevenness the only item in the sandwich that came under scrutiny, although enthusiasm for Expamet won the day. A number of other ready-made patterned and dimpled stainless steel sheets were seriously looked at.

The main contenders were from a range of decorative wall panelling by Rigidized Metals Corporation, Buffalo, New York, USA, in 302 stainless steel with satin finish, or bright annealed. Rigid-tex pattern 1-ACC (Fig. 14) with an actual thickness of 0·035 in, sectional thickness 0·135 in when rolled, had very similar transverse grooving to Expamet pattern 197s and was very much stiffer and able to withstand

(a) (b)

FIG. 14. Alternative inserts to expanded metal.

repeated inconsiderate usage. It was much easier to keep clean, but did not have the perforations which not only gave the expanded metal sheet 'teeth' but allowed the escaping vapour access to both sides and therefore more grooves.

Rigid-tex pattern 2-FL (Fig. 14) with an actual thickness of 0·035 in, sectional thickness 0·075 in when rolled, had possibility as a light-weight sturdy tray, its contour perhaps matching an identical plate forming the skin of the heating labyrinth assembly. At a later date it was thought of as having potential in the fabrication of a finned dish, but again aluminium extruded sections were tried first and lent themselves more readily to mass production. Expanded metal sheet does not take kindly to rough handling, distorting easily, although straightening just as simply; there is still some scope for ingenious thinking along such lines.

Chapter 4

The Present Position

Competition is just as keen amongst food freeze-drying firms as it is in other sections of technological industry, and reticence of mutual awareness just as prevalent on matters of apparent individual company 'know-how'. Consequently, while information in this chapter may appear general, it has been assembled in good faith. Likewise, some has been released in limiting confidence, and is not here as explicit as one perhaps might prefer, yet ample in the context.

Only a few 'thumbnail sketches' are provided to illustrate the diversity of company profiles involved, probably more multi-form than the range of freeze-dried items currently available to knowledgeable buyers. As at 1984, the following Table 1 represents production freeze-drying plants according to the legend of freeze-dryer manufacturers.

Freeze-drying of foodstuffs *in* England today is by contract and virtually dependent on a private company, Commercial Freezedrying, at Clitheroe, Lancashire. This venture started in 1968 by having its first machine built from unrelated equipment adapted to suit. Thus the tandem drying chambers were linked to a common ice-condenser chamber, all three cylindrical vessels being ex-use vegetable chillers, already proven as capable of vacuum work. Electric heating was by resistance meander wiring under simple shelves on which sat deep man-handleable trays. Control was in single steps and there was no tray cooling means. The refrigeration equipment and the pull-down/maintaining vacuum pumps were converted mismatches.

This *ad hoc* assemblage *worked*; the disparities did not really matter, because such advantageous purchases and determined linkages brought business immediately, sufficient to enable the dedicated

TABLE 1
Production Freeze-Drying Plants (Foodstuffs) as at 1984

Africa	*Europe*
Angola: 2Acr	Belgium: 2L4$\frac{1}{2}$t
	Denmark: 5Acr
Asia	Eire: 4Vca, 2Scr, 1Aca
India: 3Acr	France: 2Lc
Japan: 1Aca, 4At	Holland: 7Aca, 1Lc, 1Lt, 1Lv
Malaysia: 1Acr	Hungary: 2Lc
Pakistan: 1Lcr	Italy: 9Acr, 1At, 1L2t, 1Cc
South Korea: 2At	Norway: 2Lc
Taiwan: 8Acr, 1At	Portugal: 6Acr
	Spain: 1Acr, 1L3t
Australasia	Sweden: 3Lc
Australia: 2Lt	Switzerland: 3Acr, 1At, 1Lv
New Zealand: 1Aca, ?Cc	UK: 3Aca, 3Cc, 3Ec, 1L2t, 1Vt, 4Vca
	USSR: 3Aca
	West Germany: 1Aca, 5Acr, 3Lc, 7L4t, 1L3Jt, 4L2t, 2Lv
	North America
	Canada: 4Sc, 2Cc
	USA: 14Fc, 2Aca, 2Cc, 23Sc

A, Atlas; L, Leybold; F, FMC; S, Stokes; V, Vickers; E, Edwards; C, custom built; a, accelerated; c, cabinet; t, continuous; J, 3 m diameter; r, radiant heating.

returned-expatriate directors to build capital steadily from which to rationalise their production.

A number of prominent food processors, specialising in convenience foods, themselves unwilling to indulge in the direct expense of purchasing orthodox plant, or perhaps feeling unable individually to sustain sufficient production, became regular customers. They delivered material in bulk, ready prepared and individually quick-frozen, and collected the resulting product also in bulk.

Initially input was 750 kg, loaded at −20°C and dried for 21 h on a 24-h routine. Cabinet pressure ranged between 0·8 and 0·5 mm Hg; prawns and chicken pieces were dried concurrently. Vacuum release with N_2 gave half an hour comfortably to pack the produce in triple-wall sacks, also furnished.

A subsequent machine had liquid circuit heating/cooling/freezing built in, being used to dry 50 kg loads of fruit pieces. The control system concept for the entire installation operation was next re-examined, and rendered more effective (and less demanding of prolonged operator manual-setting) by programming the heating cycles into computer memory chips. The latest extensions at this site include a virtual doubling of capacity in the form of yet another vacuum chamber embodying further enhancements to procedure derived from 15 years of catering to exacting demands. Temperature sensing is by spade thermocouples inserted in the deep beds, and almost stepless control of the electrical heating is now achievable.

Within the past year a second contract freeze-drying factory has been installed at Preston, some 20 miles west, in collaboration with an American organisation. The layout, processing and personnel arrangements are designed to meet EEC quality approval. Capacity on prawns, chicken pieces and strawberries, for instance, is thereby doubled. Autec Lyovac chambers continue in favour, and automatic control of each cycle is further refined.

This success story bears the hallmark of realising a need, approaching a commercial solution determinedly but with common sense, accepting the stringent quality control restrictions inevitable in such a venture, and receiving the sustained backing correspondingly deserved.

Sustenance of the casual camper, the hill-walker, the mountaineer in Britain generally, as well as numerous expeditions to Everest, K2 and the Poles, has been met in major proportion by a young Leicestershire company (Haldane Foods Ltd, now partnered by British Arkady) who assemble a range of quickly reconstitutable meals from freeze-dried foodstuff processors at home and overseas. An indication of their mixes is given in Table 2. Given that their basic ingredients are delivered in bulk, their special skills tend to be associated with compilation, combination and final retail packaging in a controlled working atmosphere which is designed to prevent undue moisture pick-up, but still permits of comfortable working conditions, all towards an acceptable finished goods storage life.

During 1988 James Finlay plc of Glasgow, Scotland, who for many years have been major producers of black tea and of spray-dried instant tea (from Kenya), installed three 72-tray freeze-driers in their Hull premises to make freeze-dried instant tea. The product is reported to be of excellent quality, being noticeably better than spray-dried instant tea. It will be sold in bulk for repacking by customers in the retail, vending and catering trades.

TABLE 2
Typical Camper Snacks Reconstituted by Adding Boiling Water to the Original Sachet

Twin packs
Sachet of freeze-dried rice with sachet of
FD chicken
FD beef in burgundy
FD beef, curried
FD paella

Single-sachet pack
Freeze-dried { potato with beef hash
potato with vegetable hot-pot }

FD spaghetti bolognese
FD beef with semolina and herbs
FD chicken and mushroom
FD beans

Freeze-dried banana with custard powder
Freeze-dried apple with custard powder
Freeze-dried fruit cocktail (apple and strawberry)

The 1989 expedition of Sir Ranulph Fiennes' further attempt *unsupported* on the North Pole, at the time of writing delayed until next year, as previously will carry a food supply from this company in which they have used freeze-drying techniques to put the products together.

The New Zealand story hinges on three entrepreneurs, one the innovative industrial chemist of a South Island meat freezing works, who, exhilarated at prospects of lightweight compact meat shipments from his hemisphere to the other, persuaded an importer/exporter to bring an early AFD plant into South Island for his freezing company, and the second a self-taught engineer who started building prototype shelf freeze-driers there for a smaller but equally enterprising company specialising in frozen vegetables. There was ample supply of suitable cooling water for such operations but the steam-actuated vacuum system for the large imported cabinet tended to deny that factory an appreciable amount of its boiler output. Instead of proceeding to reinforce their steam-raising capacity, after a pause they had the freeze-drier converted to refrigerated condenser and vacuum pump working. In that pause, Dobson Foods had installed 8 orthodox freeze-driers, each of 23 m^2 shelf area. From these encouraging leaps, the versatile pioneering firm of W. G. G. Cuddon, remote from competition or indeed guidance other than from Massey University, resolutely set about designing and manufacturing their own range of high-quality industrial driers as well as laboratory and pharmaceutical models. By 1980, Alliance Freezing had added another 4 cabinets and 200 m^2 to their capacity.

Cuddon's custom-building efforts were many and varied, but a typical late design, 55 m^2 in shelf area, in widespread use in Australasia, is shown in Fig. 28. The cabinet is a cylindrical mild-steel shell 2·2 m long and 1·8 m in diameter, epoxy resin painted, with a door at each end; 105 aluminium trays 1 m × 0·54 m sit on two banks of hollow, hot-water heated, stainless steel shelves, 50 mm spacing, which can be tilted for easier cleaning, on occasion uncoupled and rolled out on rails. The vapour condenser, protected by a baffle-plate, is an eight-circuit evaporator on R502 from a 45 kW compressor, lies beneath the shelves and surrounds the take-off to the 70 dm^3/s gas-ballasted Edwards 11 kW vacuum pump. Alliance Freezing (Southland) have five such plants producing 3 tonnes per week of freeze-dried foods and freeze-dried haemoglobin, various glands and mussels in powder form for the camping supplies market and the health food industry respectively, with bulk delivery of freeze-dried cooked meats, vegetables and seafoods to food processors for preparation of gourmet soups.

The Meat Industry Research Institute of New Zealand report that tariff barriers of 20% duty on ovine and 26% duty on bovine products entering the EEC make it difficult for New Zealand to compete with advantage in this process for its traditional customer nations.

The third entrepreneur was an Auckland trawlerman whose firm now specialises in freeze-drying raw green-lipped mussels, at one time the staple diet of the Maori. Since these people started eating Western foods the incidence of arthritic disorders amongst them was noticed to have increased. He sent some for analysis to US researchers screening different types of shellfish. Clinical tests and double-blind trials seemed to confirm that an extract was as effective as standard treatments in relieving some symptoms. Research at the University of Auckland has revealed that, rather like aspirin, the mussel extract inhibits biosynthesis of prostaglandins, compounds of fatty acids which contribute to arthritic inflammation. Be that as it may, freeze-drying in New Zealand from 1978 thereby had an expanding export market, and is able to sustain a marketing as well as a processing industry.

Oregon Freeze Dry Foods Inc. has been supplying US government agencies regularly for 25 years, a factor which undoubtedly has secured its continuity. Owned by the Seven-Up company, part of the Philip Morris group, its range is marketed under the Mountain House label. Besides providing substantial amounts of ingredients for own-label soup mixes, and long-range patrol rations for the US Marines, the company has succeeded competitively in supplying a wide range of items in bulk to the food industry for use variously in salad dressings, cake, yoghurt and dessert mixes, and health food formulations.

A particular market line provides 'true flavour and true natural colour' of peaches, strawberries, oranges, lemons, pineapples, raspberries and kiwi fruit, freeze-dried into powders, granules or as slices, for use in ice-creams, fondants and chocolate-enrobed cream-centred confections.

Another angle, directed towards the idea of standby stocks for the family, provides (as a purchasable unit) up to 120 days supply of each of three balanced meals for one average adult, the menus being dietician planned. The 108 (3 litre) cans are nitrogen packed but with re-sealable lids.†

In furtherance of their claims of reliability, a long-term storage test at 21°C constant, initiated in 1975, continues to demonstrate (in proper packaging) product shelf stability. Storage reaching 33°C, such as can happen in the Americas, is not recommended for full retention of vitamin and flavour quality.

The Suwelack family company in Westfalia, some 90 years ago the largest dairy in Germany, was exporting spray-dried milk powder before the First World War, and far-sightedly installed its first tunnel freeze-drier (of 92 m^2 heating surface) in 1963. By the mid-1970s it had outgrown its original premises. The preponderance of activity in the new works (while retaining interest in several forms of drying) gradually moved from dietetic and general convenience products to freeze-drying, principally of meat, dough, fruits and beverage components. The latter became a key expansion factor in the contract drying side of their business.

† Foil packs with nitrogen in the head-space are featured for campers and explorers with the proviso of use below 3 km of altitude. There the atmospheric pressure would be less than 70% of that at sea level. For higher venturing, the pouch should be 'pricked with a pin, then taped, and used preferably within 2–3 weeks thereafter' (surely an elaborate precaution in extremity!).

TABLE 3
Freeze-Drying Plants on Which Information was Available as at 1985

Processor	*Location*	*Code*	*Maker*	*Type*	*Year*	*Products*	*Remarks*
Abels	Auckland, NZ						
Aerated Foods	Aberdeen, Scotland	1Aca	Atlas	1/3AFD		Fish, tea	Extinct
Agro-FIL Coop.	Catania, Sicily	3Acr	Atlas	3 Ray 75		Orange juice	
		1At		1 Conrad 200			
Alliance Freezing	Invercargill, NZ	4c	Cuddon	3@500, 1@600 ft^2			
Armour	Chicago	2Aca	Atlas/Stokes	2 AFD Mk 1a		Meat	
Asia Food Corp.	Taipei, Taiwan		Atlas	2 Ray 75C			
Australian Egg Mktg Bd			Leybold	2 × 1 tunnel	1968	Egg prod.	Now in W. Germany
Austral. Frs Res. Est.	Scottsdale, Tasmania	2c	Budge	2 cyclic			
Blanchaud	Mne-et-L, France					Mushrooms	
Brooke-Bond			Vickers	Tunnel		Herbs/fish white	See Unilever
Campbell Soups	New Jersey, USA						
CocaCola Industries	Brazil		Leybold	4 sect. tunnel	1969	Coffee extr.	Absorpt. refrig.
Cocam SA	Brazil		Leybold	5 sect. tunnel	1970	Coffee extr.	Absorpt. refrig.
Commercial Freezedry	{Clitheroe, UK Preston, UK	3C	Autec	Cabinets	1955–80 1988	Meat, fish	
Confidential client	Germany		Leybold	G250	1972	Diet food	
Dahlenburg Mölkerei	West Germany		Leybold	4 sect. tunnel	1964	Wide range dairy	
				2 × 1 tunnel	1967	prod., veg., fruits,	
				2 × 1 tunnel	1967	coffee extr.	
				2 × 1 tunnel	1967		
				2 × 1 tunnel	1968		
Dalhoff & King	New Zealand		Atlas	1 AFD Mk 1a modified			Agent for Alliance
Danish Freezedrying	KirkeHyllinge, Denmark		Atlas	1 Ray 75c		Shrimps, fruits,	
				3 Ray 75d		herbs and spices,	
						fish	
Deshidratadora de Alimentos	Arequipa, Peru		Leybold	4 sect. tunnel	1967	Garlic etc.	Absorpt. refrig.
Deutsche Extrakt Kaffee	Hamburg, FRG		Atlas?			Coffee extr.	Atlas Freezing
Domo/de Jong Verenigde	Netherlands		Leybold	Chamber	1958	Soup ingred.	
Dong Suh Foods Corp.	Seoul, S. Korea		Atlas	1 Conrad 200			Absorpt. refrig.
				1 Conrad 400			

Dragon Gate Food Corp.	Taipei, Taiwan	Atlas	2 Ray 75c			
			4 Ray 75d			
Eastern Freezedry Corp.	Lancaster, PA, USA				Mushrooms, poultry	Hanover Foods
					200 t FD peas/year	also IQF veg.
EDEL	Liege, Belgium	Leybold	2000 t/y	1979	Coffee	Trademark: Armada
Erin Foods	Midleton, Eire	Stokes	200 m^2 deep tray			
Faema	Milan, Italy	Atlas	2 Ray 75c			
			4 Ray 75d			
Fed. Nacional de Cafeteros	Colombia	Leybold	5 sect. tunnel	1979–80	Coffee extr.	Absorpt. refrig.
Fine Foods International	Berlin, Germany	Kraus-Maffei			Coffee	
Frozen Foods VACRS	Gaoxiong, Taiwan	Atlas	1 Conrad 200			
Frutas Soluveis FRUSOL	São Paulo, Brazil	Atlas	1 AFD Mk 2			CocaCola
Faema Cafe	Barcelona, Spain	Atlas	2 Ray 75d			
James Finlay plc	Hull, England	Edwards	3 cabinets	1988	Tea extr.	Some contract drying
General Foods Ajinamoto	Suzuka, Japan	Atlas	4 Conrad		Wide range	
Gen. Fds Calif. Veg. Concs	Modesto, CA, USA			pre-1966	Wide range	
Gen. Fds HAG	Bremen, FRG	Leybold	4 sect. tunnel	1966	Coffee extr.	Abs. refrig.
Gen. Fds	Heilbronn, FRG	Leybold	3 sect. tunnel	1969	Coffee extr.	3 m diam abs. refrig.
General Foods Post	Battle Creek, Mich., USA			pre-1966	Fruits	
HACO	Gumligen, Switzerland	Atlas	3 Ray 75d		Coffee	
			1 Conrad 200			2 absorpt. refrig.
Chr. Hansens Laboratorium	Copenhagen, Denmark	Atlas	1 Ray 75			
Heinz/Rossi & Catelli	Palma, Italy	R & C			Soup ingred. (chives, spring onions)	
Hardie Fms/Freezedry Fds	Oakville, Ont., Canada	Custom-built	2 @ 242 m^2		Soup ingred.	
CVF/Incataca	Rubio, Venezuela	Atlas	1 Conrad 300		Coffee	2 absorpt. refrig.
Indian Government	India, 3 sites	Atlas	3 Ray 75c			
Innovative Foods	Los Angeles, USA				Fruits, chives, ice-cream	
Ishigaki Foods	Tokyo, Japan					No trace
Jacob	Bremen, FRG	Leybold	4 sect. tunnel		Coffee	Extinct (absorpt.)
		Kraus-Maffei	Vertical			
Jiffy Foods (Nippon Jiffy)	Kyoto, Japan	Atlas	1 AFD Mk 2			No trace
KABI	Stockholm, Sweden	Leybold	GT20, 30, 75	1972		

TABLE 3—*contd.*

Processor	*Location*	*Code*	*Maker*	*Type*	*Year*	*Products*	*Remarks*
Knorr (CPC)	Heilbronn, FRG					Soup ingred.	Extinct
P. Kolln Flockenwerke	Elmshorn, FRG		Leybold	GT10	1971		
Liangol	Luanda, Angola		Atlas	2 Ray 150d			
Liebigs Extract of Meat	Sligo, Eire		Atlas	AFD Mk 1a		Meat/fish	Dehy Meat/Unilever
Liobras	São Paulo, Brazil						
Liodietetici	Rome, Italy						Atlas Freezing
Liogal Liofilização Portug.	Lisbon, Portugal		Atlas	2 Ray 75			
Lioprasa	Valencia, Spain						Atlas Freezing
Lioprodotti	Promezia, Italy						Atlas Freezing
Liobel	Belgium		Leybold	2 × 4½ sect. tunnel	1975	Coffee extr.	Absorpt. refrig.
Lyons/Sol cafe	London		Leybold	2 sect. tunnel	1968	Tea/coffee	Extinct
Lyons Tea	London		Atlas	AFD Mk 1a			From Aberdeen
Cia Mogi de Cafe Soluvel	São Paulo, Brazil						4 Atlas freeze. lines
MacFarlanes Lab.	Auckland, NZ	3c	Cuddon	2@600, 1@300 ft^2		Gr mussel	
Mafco	Kuala Lumpur, Malaysia		Atlas	1 Ray 75d			
Medicar Werke	Budapest, Hungary		Leybold	GT15	1973		
Mellin d'Italia	Milan, Italy				pre-1966	Baby foods	
Mölkerei Dahlenburg	West Germany		Atlas	4 Ray 75d			Coop. absorpt. refrig.
Nestlé/Afico	West Germany		Leybold		1958	Soups	
Nestlé	Auckland, NZ						
Nestlé/Maggi	Singen, FRG		Atlas	AFD Mk 1a		Soups	
Nestlé	Mainz, FRG					Coffee	
Nestlé/R. Hollandia Mlk Fd	Vlaardingen, Holland		Atlas	6 AFD Mk 1a			
Nestlé/Sitpa	Auxonne, France		Leybold		1959	Soups	
Nestlé/Spain	Spain		Leybold	3 sect. tunnel	1969	Coffee extr.	
Nyegaard	Oslo, Norway		Leybold	GT32, 150	1971		
Ontario Food Products	New York, USA		Stokes				Ex Lipton
OPDO Gesellschaft	Suchtein, FRG		Leybold			Meat	No trace
Oregon Freezedry	Albany, Oregon, USA		FMC	14-1t		Wide range products	7 Up/Ph. Morris
Ortlieb	Bensheim, FRG					Veg.	No trace

Pakistan Foods	Chittagong, Bangladesh	Atlas	1 Ray 25			
Preservenbedrijf	Breda, Holland	Atlas	VCD, converted			
RHM (Affined Foods)	High Wycombe, UK	Own make			Egg	Atlas Freezing
Right Away Foods	Edinburgh, Texas, USA	Stokes			Scrambled egg, prawns, steak, compressed veg.	Ex United Fruit
SACAF(F)	Salerno, Italy	Leybold	2 sect. tunnel	1962	Fruit/veg.	Now coffee
Sec-Aliment	Barcelona, Spain	Atlas	4 Ray 75d			
Soc. SERCAL	St Cyr, France					Atlas Freezing
Slack Bros	Quebec, Canada				Mushrooms	
SOLIS	Sardinia				Coffee	Atlas Freezing
Solubles Instantaneos	Guayaquil, Ecuador	Atlas	2 Conrad 300			
SPAB	Saumur, France				Mushrooms	
Südmilch	West Germany				Soup ingred.	CPC supplier
Dr Otto Suwelack	West Germany	Leybold	2 sect. tunnel	1964	Wide range	Absorpt. refrig.
			4 sect. tunnel	1967		
			4 sect. tunnel	1968		
			4 sect. tunnel	1969		
			1 vertical vibrator			
Technopromimport	Moscow, USSR	Atlas	3 AFD Mk 1a			
UngarischesKalteinstitut	Hungary	Leybold	G100	1965	Fruit/veg./coffee	
Unilever/Batchelor Foods	Ashford, UK	Vickers	AFD			Extinct
Unilever/Bla Band	Halmstad, Sweden					
Unilever/Dehy meats	Sligo, Eire	Unilever				See Liebig
Unilever/Hartogs	Oss, Netherlands	Leybold				
Unilever/Lucas Aardenburg	Netherlands	Leybold	1 tunnel	1964	Vegetables	
Urform	Herrenberg, FRG	Atlas	1 Ray 75d			
Warnick/Hirschberg	Burlingame, CA, USA	Stokes				
Watties	Gisbone, NZ					
Wilson	Omaha, Neb., USA					

Liquors prepared by customers to their individual formulation are delivered by tanker to Billerbeck to be frozen, fragmented, freeze-dried and bulk-packed. Import/export regulations have accommodated the subsequent final retail-container (re)packing in the customer's country. Expansion has on occasion included acquisition of (second-hand) freeze-drying tunnels from Australia, just to maintain the impetus.

While coffees have become a mainstay of this centre, Suwelack is also a leading exponent of freeze-dried collagen derivatives for skin-grafting and wound dressing.

A summary, complete as far as could be obtained, of freeze-drying plant details by processor is given in Table 3.

Part II

THE FREEZE-DRYING PROCESS

Chapter 5

Aspects of Foodstuffs

Unlike the koala's eucalyptus leaves, or the panda's bamboo shoots, mankind's preferred diet is utterly diverse. There are of course, climatically, staple ingredients such as potato, maize and rice, but one may still discern the sensible primitive reasonings behind selection of germ-of-life stores in eggs, fruits and nuts, and the incorporation of tubers, young plants and the muscle systems of powerful animals.

Nowadays a concerned food retailing industry, encouraged by government, prevails upon the capricious public in a developed economy to monitor their health via nutritive intention, i.e. at the point of purchase, but simultaneously there exists the possibility of being waylaid by a wealth of competitively marketed and organoleptically attractive wares

Of the wide range of commodities whose heralded arrival at distribution's terminal is aided by a preservation process, beyond the obvious consideration of equilibrium moisture content lies an inculcated suggestion that the foodstuff as presented should contain no waste, and moreover be of a familiar shape, convenient size, and in a nearly ready-to-use condition. This sets a scale of restrictions or benefits from the *processor's* point of view—familiar cuts† offering predetermined temperature gradient obstructions during freezing and drying sequences, and complications in uncontrived vapour escape

† Compared to the overall dimensions of bite-size morsels of foodstuff, individually most cells are tiny. It follows then that paths both tortuous and lengthy, and non-constant (but often averaged) rates of heat transfer out or in, are seriously dependent on the arbitrary concepts differentiating the nature of wall and connective tissue.

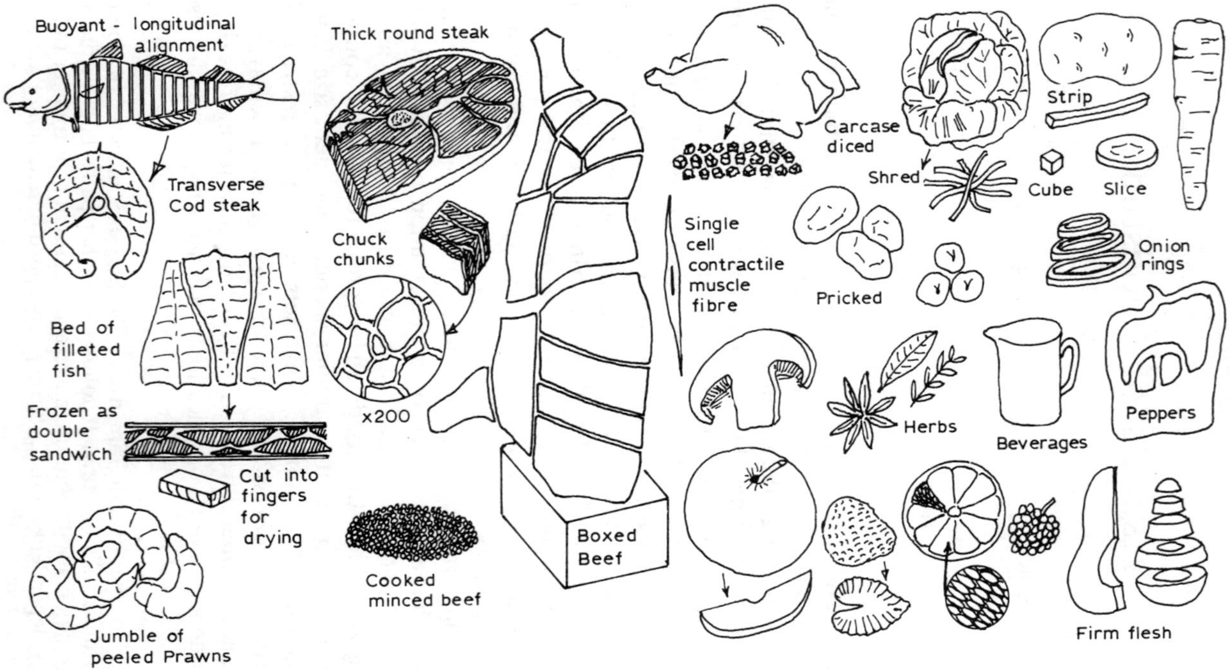

FIG. 15. Forms for freeze-drying. Most foodstuffs are prepared for drying as conveniently sized pieces, familiar and attractive, but randomly oriented. Consequently, slopes of gradients (of both heat and vapour transfer), within such a variety of textures, beds and pathways, are difficult to predict universally by mathematical treatment.

paths (Fig. 15). Reconstitution time, which has to be 'reasonable', is also predicated.

The distinction between a foodstuff which contains in plenty the chemical building bricks from which, on digestion, one's body can comfortably reassemble the compounds strictly to be identified as *foods,* and something calculated merely to bring out saliva, is a matter of biochemistry as well as ethos. A minute part of a plant or a part of an animal, even identical to part of another animal, will not make the transition to become a true part of that other living creature without complex sub-division and reassembly (Fig. 16). The procedure demands several prerequisites. The ingested material must not have suffered such harm, physically or chemically, as would defeat the purpose; it must contain, in due proportion, the formulations necessary. The body will discard perfectly good excess amino-acids if the balance is not right, but the instincts that would prompt it to direct the lack of balance into a sustained and discerning search or craving to redress such a (perhaps temporary) omission seem largely to have

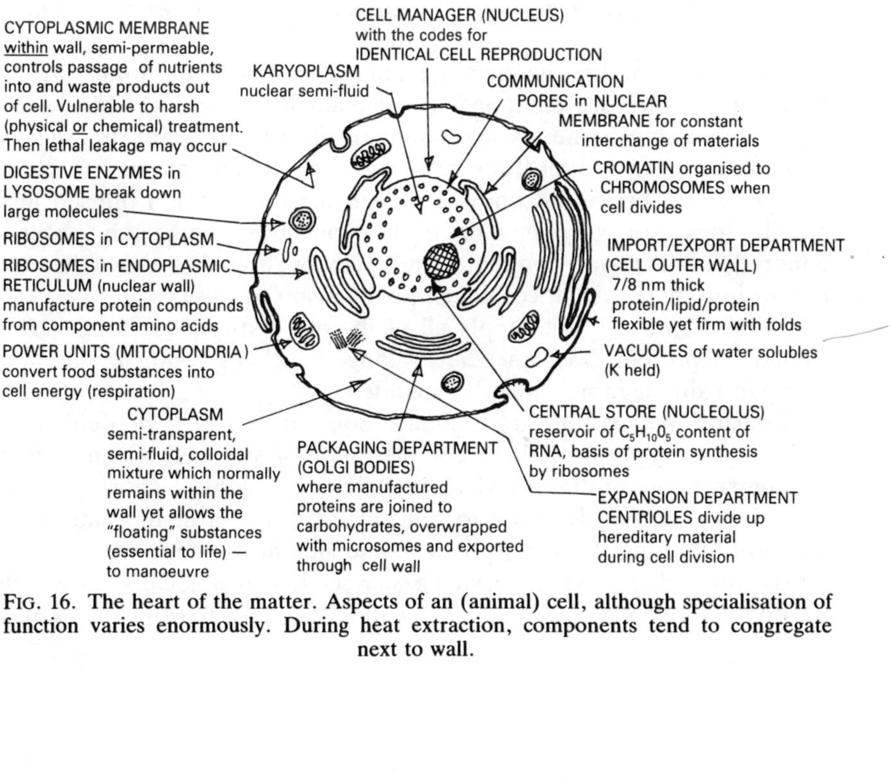

FIG. 16. The heart of the matter. Aspects of an (animal) cell, although specialisation of function varies enormously. During heat extraction, components tend to congregate next to wall.

disappeared from the innate memory. All that may remain is a desire to *over-eat,* i.e. to increase the throughput, and make gobbledygook of a diet-sheet. Such behaviour is not fully understood.

The life cycle elements in the human body to be nurtured are few (oxygen, carbon, hydrogen and nitrogen make up 96½%) but they combine in Nature with themselves and other elements to present a bewildering array of compounds which would remain lifeless were it not for the manufacturing photosynthesis by green plants converting them into organic *foods* on which other forms of life depend. (The term 'foods' here is intended to be precise; 'food' is a popular generic term while 'foods' are to be thought of as basic units assimilated from 'foodstuffs'.)

The nurture of such plants requires, besides the energy of light, cyclic access to fertilising media which are either salts directly soluble in water, or become ionic through biological and other processes in the soil. Plant welfare stems from ionic balance in the solvent (pH), and this, taking account of electrical resistivity, gives a comprehensive integral which, together with distribution ratio of all ions, will portray the state of vital activity. (Interest in ionic concentrations continues throughout processing; resistivity is measured during freezing to gauge progress.)

Each organism has its own systematic metabolism. Some higher forms of creature are equipped to eat plant forms direct, while some ingest by preying on herbivorous animals. The hunters are themselves hunted. Man can use all three routes as second-hand sources to build and power his brain and body. He even harbours within himself less mobile forms of life, as aids in his foodstuff conversion system. Kindred lower forms complete the cycle by breaking down 'waste' matter into inert substances which plants again can use.

Cereals alone will not supply all of the one-sixth of our food which should be protein, so somewhere along the chain young plants and nitrogen-fixing legumes must be included.

Foodstuff processing, from man's point of view, is a deliberate artificial interruption of these complex natural systems, which at first he contrived to ensure survival in times of potential shortage, but nowadays it is also designed to fulfil his satisfaction in the emotional pleasure of eating. Both purposes should steadfastly be borne in mind by the technologist when servicing any section of a society, although profound knowledge of the chemical and other processes involved may tend to be dismissed as non-immediate, if not irrelevant, by the regular consumer.

CUTTING THE PROBLEM DOWN TO SIZE!

Preparation of solid foodstuffs by washing, sorting, trimming, dimensioning, enzymic inactivation by scalding, and so on, are common procedures on well known standardised equipment and need not be dwelt on here, except that *speed,* or rather avoidance of interim delays (which might adversely affect quality), coupled with *accurate* thicknessing (so that drying is as uniform as possible) are especially important for this process. Procedures particular to freeze-drying of food and drink are highlighted in the ensuing pages of this chapter.

INTERESTING PREPARATION TECHNIQUES

Spray-tower freezers, where liquid concentrates are dispersed by a secondary fluid in a refrigerated free-fall chamber, to form prills or beads in the manner of lead-shot, for subsequent freeze-drying, have been patented. Mainly to aid the preparation of coffee or tea extracts, they have not achieved widespread popularity. Prior to drying by various means, the puncturing by needles or the slitting by knives of the integument of peas has been known and practised for some time. A similar action on blueberries which have been slow-frozen, as a step prior to freeze-drying, has been claimed to aid rehydration with cream, as well as speed actual drying.

Sliced peaches, pears and strawberries, frozen to −6°C or so, have been coated with a 6% sweetening solution which did not glaze, but did cause melting. This deliberate lowering of the freezing point by using a diffusant which became an integral part of the fruit meant that the fruit was then re-frozen at −23°C prior to freeze-drying. On rehydration (by milk at breakfast) the slices were uniformly sweet.

Alternatively, such sweetening has been fostered by spray application *after* freeze-drying, and before warming off to a stable moisture content again.

Purees made up of fruit mashed with one-third by weight of sugar, blended into cream in equal part, have been 'frozen' at −23°C prior to evaporative freezing to −29°C in the drier, at 0·3–0·5 mm Hg. The result was a stable centre for incorporation in chewy confection bars. Many similar concoctions, having pudding/ice-cream/sour cream *powder* bases, have been devised.

INITIAL SEPARATION (BY FREEZING) OF THE 'WATER' CONTENT

The schematic visualisation of cellular structures for the purpose of studying this process is suggested in Fig. 17. Briefly, from the freezing point of view, amid a matrix of cellulose fibres, there are plastids which harbour gel formations and vacuoles which harbour water solutes.

The common solidification concept for food pieces is that of simple metallurgical eutectics, in which the chemical components are completely and collectively soluble in a common liquid (based on water) and entirely insoluble in the solid state (based on crystals, large or small, of pure ice). When the moisture molecules have been frozen out of solution, as far as possible by progressive concentration of the 'alloys', there remain composites which refuse to form crystals or take part in crystalline structuring.

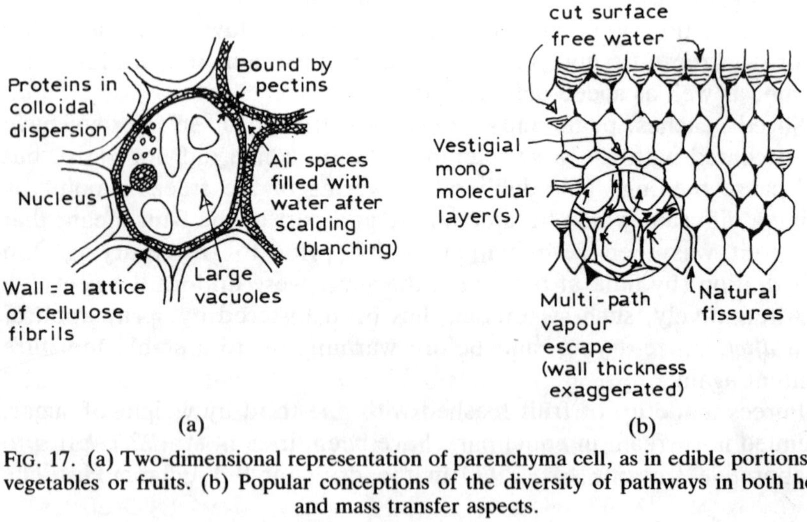

FIG. 17. (a) Two-dimensional representation of parenchyma cell, as in edible portions of vegetables or fruits. (b) Popular conceptions of the diversity of pathways in both heat and mass transfer aspects.

If the whole mass is plunged in temperature, then a vitreous or amorphous condition is obtained throughout. This state can be resolved by bringing the temperature back up again, gently, until at some special temperature a *eutectic* is reached where *both* (types of) component(s), in fixed proportion to each other, have crystalline structure, and by definition would melt simultaneously if the temperture were allowed to rise further.

This hesitation or dwell point is identified by a pronounced change in electrical resistivity. Strictly, when the solid solution breaks up into *two* intimately mixed constituents, they form a *eutectoid,* and the temperature at which this just occurs (the melting point) will be the highest temperature at which true freeze-drying may be carried out. The second-order phase transition will occur at a somewhat lower level of temperature.

The actual conglomeration of protoplasmic substances is not totally sub-identifiable. Certain known components, when tested individually in model systems, show clear differences in thawing levels, such that make it odd to relate them in joint behaviour, unless some other factor is present. Much is conjecture.

Extending the metallurgical analogy still further, if reaction(s) were to take place, at definite temperature, between proportions of the non-ice components, whose concentration has been 'upset' to such an extent as to make their harmonious and functional coexistence less durable, but their activity towards each other more defensibly attractive, then this would offer an explanation not only as to why some food characteristics apparently alter in the 'frozen' state, but also why living cells die in such protective conditions.

Investigation of possible peritectic reactions is insufficiently reported.

VIGILANCE

The general public is still only vaguely prepared to admit that, whether they like the idea or not, they spend their lives in association with trillions of micro-organisms whose survival and growth are favoured by conditions in which they themselves normally fulfil their lives. Fortunately most are harmless, an invasion by those potentially harmful being opposed by natural defence (and chemotherapeutic) systems. However, since our surfaces and body openings are typical host sites to the 'parasites' category, and we tend to share our choice of foods with the numerous 'saprophytes', it behoves us to be particularly and thoroughly careful in personal and equipment hygiene when involved in the preparation of food for others.

On rough average, micro-organisms regenerate some 130 000 times faster than humans, so colonies of them can quickly get out of hand. (The pun may be forgiveable!)

A guide to good manufacturing practice is published by the Institute of Food Science and Technology (UK) which points out the strict discipline required of the industry (Table 4).

It should be noted that, while every care may be taken to keep microbiological spore loads to a microscopic minimum (and the record shows such success), even thermostabilised foods revert to being vulnerable once their protective packaging is removed. Freezing of itself, while not actually killing decay bacteria, arrests their growth and prevents them from spreading dangerously. Indeed, dried materials may well present slower deleterious response when exposed (after packaged storage) to a hostile environment. Such a conscious act is subject to the common sense acquired from habitual culinary observation and usage. One cannot be too careful!

The target of removing, by 1992, *all* trade barriers within the European Economic Community is of special concern to the foodstuffs industry, all twenty or so facets, where national tastes and zone characteristics have entrenched meaning, where the Common Agricultural Policy is hardly in universal favour, and where consumer lobbies point to a health and social as well as an economic dimension.

During 1987 the EEC sold 3/4 million tonnes of butter to the USSR, cutting its butter mountain by half. Attempts to dent the stockpiling costs of the previous year by concentrating butter, i.e. reducing the normal 15% or so of moisture content by evaporative drying under vacuum to 1–2%, making it more suited to shallow frying and basting, and 20% more economical for baking, had not found much favour with the general public of the various member countries. It did, however, indicate that technical means of reducing the cost of keeping food commodities in long-term storage was being considered by the authorities, and that they were prepared to employ marketing publicity to introduce such variations to the populace at large.

The enzymatically modified starch β-cyclodextrin was approved in the EEC during 1986–7 for food use as a flavour carrier. Claims for CDs generally include: with gaseous guest molecules aromatic vola-

TABLE 4

An Indication of the Stringent Requirements Imposed on a Dry Soup Mix Ingredient

SAMPLE PRODUCT SPECIFICATION
DATE OF ISSUE: January 1985

PROCESS: FREEZE-DRYING
PRODUCT: GARDEN PEAS
SIZE: FLAKES

Physical requirements	SIZE: 100% will pass through a BSS No. 4 Sieve (4000 microns) Not more than 10% will pass through a BSS No. 36 (420 microns) 100% held in a BSS 100 (150 microns)		
Sensory requirements	When rehydrated, the pea flakes will have the colour, flavour and texture of fresh peas and shall be free from foreign flavours and odours.		
Defective tolerance	There should be no apparent defects in the product and foreign material shall be absent.		
Analytical data	(a) MOISTURE CONTENT:	Maximum 4·00%	
	(b) SO_2 CONTENT:	300/800 ppm (Target: 500 ppm)	
	(c) ENZYME ACTIVITY:	Peroxidase Negative	
	Note: Sulphite can be excluded if so requested prior to harvest.		
Bacteria	(A) TOTAL VIABLE COUNT:	Less than 100 000 per (dry) grm	
	(B) TOTAL COLIFORMS:	Less than 100 per grm	
	(C) *E. COLI*:	Normally Nil (Max 10)/grm	
	(D) SALMONELLA:	Absent in 25 grm	
	(E) *STAPH. AUREUS*:	Less than 100 per grm	
	(F) *CL. PERFRINGENS*:	Less than 10 per grm	
Packaging	Polythene lined Fibre Board Cases (Specification No. E 200)		
Pesticides, heavy metals		MAX	mg/kg
	Benomyl, Carbendaxim	Total	0·3
	Bromid	—	50
	Chlordan (A, Y-Isomer)	T	0·1
	Chlorpyrifos	—	0·1
	DDT/METAB, (O, P′-+P, P′-ISOM)	T	0·3
	Dichlorvos	—	2
	Dieldrin, Aldrin	T	0·02
	Dithiocarbamate (AS C32)	T	0·1
	Endrin	—	0·005
	Heptachlor, Epoxid	T	0·02
	Hexac Ilorbenzol	—	0·05
	Hexachlorcyclohexan, Y-Isomer	—	0·2
	Malathion	—	8
	Parathion, -Methyl EACH/JE	—	0·05
	Phosalon	—	0·1
	Hydrogen phosphide	—	0·1
	Pyrethrine	T	3
	Quintozen	—	0·01
	Lead	—	0·5
	Cadmium	—	0·1
	Mercury	—	0·01
	Copper	—	15

tility is reduced; liquid compounds tend to be transformed into crystalline form; oxidation is decreased; reconstitution is more rapid. Bitterness (of say citric juices) is balanced by masking. Retention of flavour components increases with higher molecular weights (of essential oils). Typical procedures are to extract from pulverised vegetables or fruits with aqueous CD solution which is then filtered, concentrated and freeze-dried to a stable flavouring powder.

Whole or ground spices suspected of microbial or other contamination have been subjected to rigorous drops in temperature as a purifying measure, followed by drying to preserve all the odour and flavour constituents. Incomplete versions obtained by distilling for essential oils or dissolving out oleoresins, to be dispersed on various bases, are considered as more standardisable, if less intense. Recognisable fragments of herbs, on the other hand, when freeze-dried, are unsurpassed in colour retention, concentrated odour and flavour.

Although particular characteristics of a foodstuff may be enhanced or acceptably changed due to processing, the quality of the raw material is the key to success (Fig. 18).

In 1980 at Billerbeck, West Germany, freeze-drying of bovine meat, which had been slaughtered only 4 hours previously, had reached 1000 tonnes per year, all in the form of ground or minced beef, free of bones and sinews, and reduced in fat. This preparation time includes salting then freezing to −40°C to −42°C in the ribbed dishes. The dried granulate weighs 70% less than fresh meat, is standardised in fat, protein and collagen content, and is principally for use in a wide variety of continental meat products, having a preferred 'warm meat' taste. The same factory produces a range of freeze-dried granulates and powders for baking mixtures from which to make cream cakes, cheesecakes and like items, all competing for shelf space on quality grounds.

This compares interestingly with the Department of Agriculture for Northern Ireland's research into improved meat tenderness in the pH range 4·6–4·1 as opposed to pH 5 where the extra loss of water during cooking makes for chewiness.

No-simmer packaged 'flash' soups have replaced slower (20 min) reconstituting (shrivelled from liquid phase drying) ingredients. Peas which are tenderometered, picked and frozen within $2\frac{1}{2}$ hours are well known. If they are to be dried they are individually pricked or scarified in a special machine to make egress easier.

A beef and vegetable stew prepared from 40% (by weight) boned

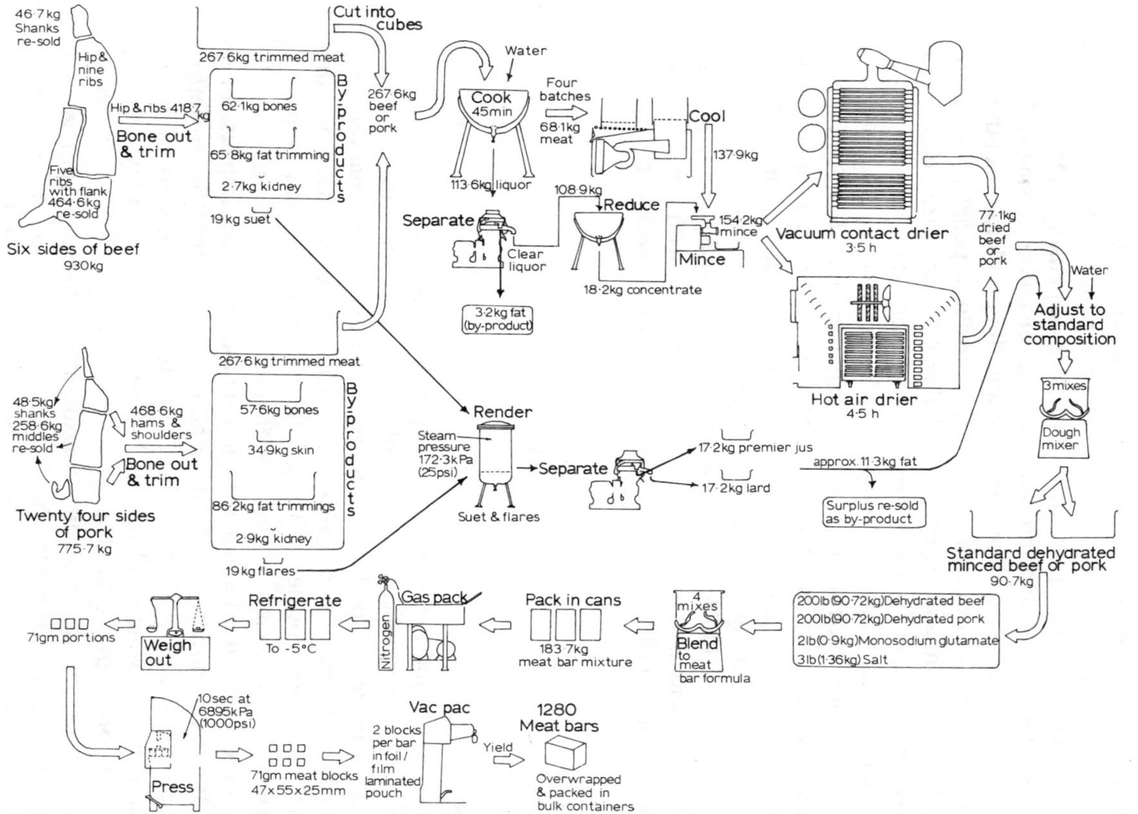

FIG. 18. The steps involved in making the original Services 'meat bar', indicating: (a) the intricate preparation, regardless of the drying method; (b) the cost effects of a rigid specification; (c) the attitude towards avoidance of oxidative rancidity.

and trimmed cubes of meat, 20% carrot slices, 17% potato dice and 12% onion rings, all amid a thick gravy, was cast by blast-freezing in loaf tins and band-sawn into 15-mm slices. The variegated, marbled appearance of the cut slab, both before and after drying between inserts, was declared most attractive and appetising. Despite the random arrangement and 'cutting through' of pieces having different textures and moisture content, which made the problem of sending the heat of sublimation through dried fibres uneven in penetration speed, overall accomplishment of drying to below 2% moisture content was reached in 11 hours (7 hours for 10 mm thick). There is little doubt, confirmed by interrupting drying on several runs to cut open and examine sample block hearts, that the expanded metal lattice on both flat sides, which dug into the slab, reducing its effective thickness by several millimetres, played a major part in overcoming the technical difficulties of limiting surface time–temperature tolerance in such a case. Cohesion as blocks was maintained during hand-packing into foil pouches. A similar binding success was achieved with another composite: spaghetti (10%) in tomato sauce (64%) with pork cubes (26%).

BEVERAGES

Coffee

It may seem somewhat ironic that a process which sought to become popular and economically acceptable as the medium of preserving truly large pieces of foodstuff should achieve its greatest market boost from materials which had a successful history of being dried by other means into *fine* powders. These, in the main soluble solids of beverages, were found to be preferred by the public in granular form—not as the hard small seed-like grains the term implies, or as conglomerated particles made to adhere to each other by re-wetting and secondary drying, but as screened sizeable fragments of frozen sheets of liquors which still retained some structural stability on being rendered porous by freeze-drying.

Apart from any other quality consideration, and there are several, the simple heaped spoonful habitual unit of *ad hoc* measurement, which helps to ensure that instantly reconstitutable beverages are dispensed both accurately enough to avoid accidental weak or over-strong infusion yet allows variation in consumer taste and preference to be judged individually, was aided by providing the dried material in grosser form.

The finer the particle size of a coloured material, the lighter its apparent colour, owing to increased reflectance, and this may colour the opinion of the observer as to its attributes. Roasted coffee extract which has been frozen indiscriminately and then freeze-dried may exhibit a light-brown, tan or even mustard-like colour, which is not quite so characteristically coffee-like in appearance. This can be attended to by altering the freezing rate, or by other preparatory means, to give a particle colour which conveys to the purchaser a truer appreciation of (say) the strength characteristic of the brew to be expected. Thus anticipation of satisfaction is built into the nature of the freeze-dried product.

Coffee beans, the seeds of the coffee tree fruit, admittedly undergo a number of preliminary processing steps (Millin, D. J. & Bradford, L., 1984) before being received as a 'commodity', with all the standardisation, swings in value and comparative assessments that this term implies. From this fascinating raw (!) material, whose concoction has to be learned, knowledgeable processors undertake to eliminate the complications and prepare brews guaranteed attractive to the consumer, arresting the distinctive brew attributes in dried soluble solids for future appreciation at a fraction of the skill but with maximised convenience.

Because of variations in consumer taste regarding home-ground coffee beans, perpetuated in the brand name rather than in the blending experience of individual households, the tendency among processors of instant coffees generally has been to prepare their own concentrated liquor of soluble solids of pure coffee in bulk, have it freeze-dried by a contract processor, not necessarily in the same country, and then pack and distribute the finished product themselves again. Consequently the freeze-drying procedure for coffee liquors has become fairly standardised, and the preparation of the liquor, while following a generally accepted routine plan, contains within it the nuances of treatment which exemplify the characteristics desired to be associated with the particular final product.

Robusta green coffee beans have slightly more organic acids (and therefore caffeine) than Arabica green coffee beans, but less coffee oil, which latter may be looked on as an add-back carrier of useful volatiles. Arabica tends to be more expensive, with finer flavour, so a blend by weight for freeze-drying may be proportionately richer in

Arabica because it is worthwhile putting more expensive beans towards a process that retains more of the aroma and flavour.

All beans are carefully screened and cleaned before being admitted to the preparation procedure. During 'medium', 'full' or 'rich' roasting the sorted beans are tumbled at appropriate speed through combusted gas, at high temperature (around 200°C at the bean), making for partial decomposition, itself giving rise to the formation of many complex volatile aroma constituents, during an exposure of 5–15 minutes.

After rapid stirred cooling, the roasted beans are crunched between shearing toothed rollers to release trapped CO_2 while yet developing preferential 'taste' effects. In a further batch sequencing operation, the coffee stays put in a number of tall columns (each holding say $2\frac{1}{2}$ to $4\,m^3$). The softened water of extraction progressively passes through and up each column in turn, from the most completely extracted grounds to the freshest roasted grounds. The temperature (and pressure) of the water is high (180°C) to begin with and falls off, so that the major part of the extractable matter is freed from the fresh coffee at comparatively low temperature, ensuring a high quality extract, while the high-temperature extraction of the nearly exhausted grounds provides a high yield. The resultant extract may contain 35% soluble solids. Some designs of installation split the flow to involve an evaporative step for further concentration of the weak liquor, so that the recombined flow achieves, say, 40% before being filtered, cooled, centrifuged and chilled for tankering. A mould-inhibiting blanket of CO_2 is sometimes used.

Thus these preliminaries to the freeze-drying process establish, partly by instrumented control but still having an incorporation of expert experience, the particular aroma and flavour, while leaving the strength satisfying individual taste to be adjusted by the amount of finished product dispensed.

Fluid-bed roasting, water-spray quenching and particle size distribution control for extraction and foam avoidance are notable among the less elaborate variations on these routines over the years. The anxiety rests in getting the brew beyond contact-with-the-bean stage, while maintaining a balance between so-called beneficial and detracting attributes stemming from inseparable permutations, the references being largely subjective.

One refinement, for instance, is to freeze-concentrate a first (aromatic) draw liquor from the extraction plant and mix it with a second (non-aromatic) draw liquor which has been concentrated by low temperature evaporation, i.e. at sub-atmospheric pressure.

Another variation is first to remove volatile constituents from the roasted extract in a wiped film evaporator, condensing them for add-back in safety later, after decaffeination in a counterflow column into which dichloromethane at around (say) 70°C is sprayed. Traces of solvent in the liquor are flashed off in a second similar evaporator, while the 97% or so of caffeine previously freed from the salts of chlorogenic acid is recovered later as a pharmaceutically useful by-product.

At the freeze-drying factory cold rooms, the preferred liquor, entering at around 5°C, is frozen to a strict regimen of both temperature and time, as this chosen procedure sets the material in the condition from which it will be dried, and influences the future colour in particular.

The fastest procedure is to pick up the chilled liquor as a thin film on the surface of a refrigerated drum whose speed of rotation determines the thickness (and ice crystal size) of the sheet so formed. A doctor knife shatters off the frozen liquor, which is further granulated by shearing mill and then middle sieved to regular size. Large and small fines are returned to the feed trough.

Alternatively, the liquor is first cooled to between −6°C and −8°C in a soft-ice machine, resulting in a mash of fine crystals which in a second stage is spread out *either* on trays to be blast frozen in an air stream *or* on a stainless steel belt floating on a sequence of brine tanks. This stage, undertaken in the region −40°C to −50°C, can be varied in duration within wide limits (2–3 h *or* 15–60 min respectively) and determines whether the product remains darkish or not. It leaves deep-frozen in the region −40°C to −45°C being roughly broken before passing to the milling and sieving system.

The ice crystals of course, as an arranged size, will sublime away to leave a porous 'structure' of soluble solids whose strands were predetermined by the distances between them. It is averred that, the thicker these strands are, the more 'locked in' are the aromatics, but of course, in order to have a pathway for vapour to the outer surface of the fragment, there has to be variation in the manner in which the individual ice crystals interlock (Fig. 19).

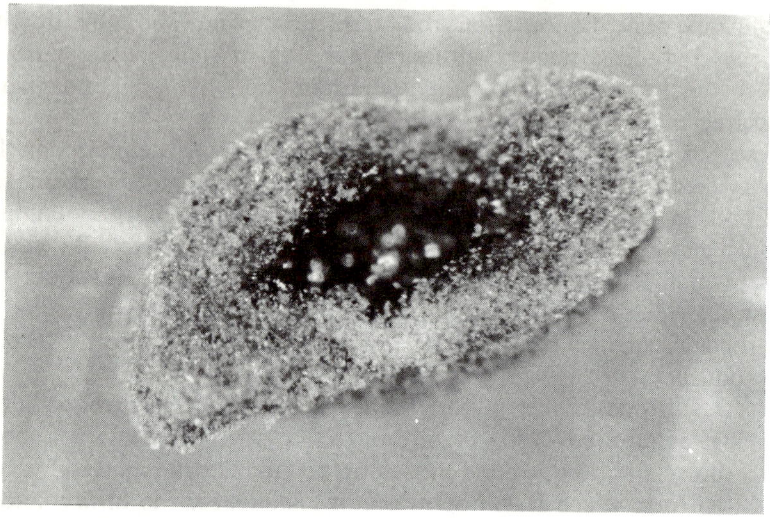

FIG. 19. Section through granule of freeze-dried soluble solids of coffee, showing hollow centre. To speed production, towards the end of a run, the heart ice has been allowed to melt into the outer (already dried) layers, from where it was easier and faster (on re-freezing) to sublime the last of the vapour away.

Tea

The preparation of 'instant' tea 'powder' is reckoned to present more difficulty than 'instant' coffee, partly because both the green and the dried black tea leaf as *basic* commodities are so much more variable (following long tradition and ritual), partly because the nuances of an infusion which might satisfy a significant proportion of the capricious public taste (whose discussion of tea in England at least rivals that of the weather) are somewhat elusive, and simply because of a practical problem—there is much less soluble solid in the brew to be isolated, captured and later, in the event, to be advisedly dispensed. More of the constituents are heat-sensitive, compared with coffee.

The character and quality-for-cost of a blend (as compared with a 'self-drinker') are of course carefully shielded against variations in origin and drying of the prepared leaf, but its highly hygroscopic nature makes it vulnerable to loss of pungency during handling, sorting, packing and storage, before it ever gets to the start of the instantising process. Preferably the moisture content should not be more than $7\frac{1}{2}\%$ (9–10% being a possible threshold for mould), and equilibrium about 13%, at normal room temperature.

Oil of bergamot is added to some tea leaf to enhance a particular aspect of flavour already there, and the citric rind containing linalool is itself freeze-dried for the purpose.

Early attempts to provide tea powder disappointed some of the eager but discerning public, who became reluctant and wary (in the face of renewed efforts to sponsor tea drinking via tea-bags for instance) of subsequent and improved endeavours. An associated complication occurred when lemon flavour crystals were added to hot-water soluble instant tea to bulk it. While making it easier not to over-dispense, the crystals produced a dominant flavour attractive to many, but tending to mask whether the tea flavour itself, with no milk added, was truly acceptable in its own right. To some extent this slowed progress.

On the other hand, cold-water soluble instant tea, well established in North America, has to provide an infusion which is transparently clear, so that the precipitate which might ordinarily form (on cooling) has to be extracted and treated to make it as soluble as the rest (some 70%) of the total solids. The final brew has also to be proof against any cloudiness or haze stemming from the use of hard water rather than locally soft.

Nevertheless, nowadays, freeze-dried 'instant' tea is available which has overcome all these problems. Processing procedures follow generally the extraction methods for coffee, with perhaps more definite cut-off points, the subsequent concentration and two-stage freezing being also closely controlled.

Tea-bags originated in the 1920s from the ancient idea that the leaves ought to be removed from the liquor (or vice versa) at the end of a specified brewing period. Various devices in metal had been contrived as recoverable infusers, leading to the emergence of the USA fine gauze cloth bag with a string and tag hanging over the side of the cup or pot. The size was either 440 or 330 to the kg of tea leaf, and the elegant practice was at some variance with the habit in Britain of topping-up from a hot-water jug, so the paper sachet type tea-bags adopted here had no string.

The normal measure per cup of instant (freeze-dried) tea powder is 0·6 g rather than 2·25 g of tea leaf, so special means (Fig. 56) had to be devised as a caution in dispensing it. Since the powder is wholly soluble, the withdrawable container serves equally as an easily disposable stirrer.

SPECIAL SITUATIONS

Explorers contemplating an arduous adventure of limited duration, such as in the Polar regions or the Himalayas, for example, are prepared to train, strengthen and equip their bodies (and minds) beforehand, so that they may draw heavily on the stored elements of the energy-summoning chain as well as their current food intake conversion, in a deliberate body-weight variation regime. Metabolic conversion rates and patterns are peculiar to individuals as well as responsive to styles of task. They have to be mapped; generalised averaging of performance is too vague. Closed-room calorimetry and isotopically labelled double-water techniques are used, with careful monitoring of carbohydrate, protein and fat, to map out a pattern of make-up and response which can be charted for at least sedentary routines. Oxygen demand is also scanned individually. Then a knowledgeable planned diet and work-out programme are used to bring the person to an optimum weight and strength before ever setting out on the expedition. Sometimes the system is checked at the foothill and/or tundra stage. The optimum sought is an increase of the energy chain base, glycogen, stored in muscle and liver, such that endurance performance is sustainable and even enhanced. Deliberately, on a regulatory diet of around 5000 kcal daily, when their work-load may be between 6000 and 7000 kcal, such intrepid adventurers may, from an 'overweight' starting condition, end up carefully calculated at (say) $3\frac{1}{2}$ kg under their 'normal' weight, but still incredibly fit.

Freeze-dried foodstuffs are a vital part of these proceedings. Property retention after processing enables analyses-based formulation to be precisely carried over. Sled or back-pack loads, initially lightweight, diminish prudently, but the water of reconstitution is obtained from ice or snow at each halt, so that variety and interest can be part of victualling without incurring excess encumbrance.

For Sir Ranulph Fiennes' great British polar quests of recent years, for instance, the caloric expenditure (of *twice* what may be reckoned as normal) was estimated as in Table 5.

The land mass around the centre of the Arctic Circle is not prominent as at the South Pole, near which constantly manned weather research bases are established, interested in aquatisation rather than sublimation phenomena. Yet in June 1987 submarines

TABLE 5
Estimation of Caloric Expenditure

Activity	*Hours*	*kcal/min*	*Daily kcal*	*Diet*
Man hauling sled	8	8·8	4224	
Rest breaks	2	1·8	216	Tomato/vegetable soup in flasks, plus chocolate bars
Making/breaking camp	2	5·3	636	
In tent, eating, cooking	4	1·8	432	Cottage pie/Bolognese/ macaroni-cheese, wholewheat cereal, sunflower oil/butter, vitamin and minerals supplement
Sleeping	8	1·2	576	
Totals	24		6084	4800 kcal in 952 g[a]

[a] The shortfall over 60 days draws acceptably on body fat.

from the USA and the UK surfaced by rendezvous amid the floes at the North Pole. Whether they exchanged FD rations was not declared.

One item in the Arctic ration is pre-concentrated, formed and sliced freeze-dried entrée in 120-g sturdy blocks. This resulted as a side-step away from compression *after* freeze-drying, vegetables being found capable of size reduction in a range 4:1 to 16:1 while dried meats remained in the range 1·5:1 to 3:1. The more effective volume reduction of the composite block was obtained by severely limiting the process water, employing microwave cooking to thicken, then forming the entrée by vacuum-assisted meat stuffer in sausage-type casings. Next it was frozen, then sliced (by bandsaw) into 18-mm thick pieces, in which condition it was dried. Sliced fruits with juice and added sugar are also frozen into 50/75 mm square blocks prior to drying followed by compacting. From the vacuum pack they are ready to eat without extra water. A not dissimilar chewy fruit bar was evolved in Aberdeen 30 years earlier, although not so compresssed.

A typical Combat Survival Ration One-Day (Wet-Pack) (code no. 276-2236) serving two light snacks and one main meal plus three hot drinks, recommended for outdoor activities like camping and soldiering, comprises a main item with sundries, as follows: 1 easy-open canister of (a) beef casserole, or (b) cottage pie, or (c) chicken casserole, or (d) beef curry and rice, to be simmered in boiling water for 15 min (of the total weight of 800 g this item is 320 g); next are 2 packets (100 g each) of 8 plain biscuits, 1 43-g canister of paté, 1 portion of jam, 2 tea-bags, 1 sachet of coffee, 3 sachets of sugar, 3 of milk substitute and 2 of salt, plus 20 matches, 1 spoon and 2 tablets each capable of purifying 1 litre of water (E224?).

A General Service 24-hour Ration Pack, averaging 1·84 kg total weight, covering breakfast, snack and main meal (in all 1200 g) in four attractive rotating menus, has the principal course items (550 g) tinned for hot or cold consumption. It relies, except for a rolled oats mix with which to make pasty porridge, on soups and beverages for use of its dehydrated ingredients. The moisture content of the pack is thus not a large proportion, but the menu supplements (270 g) involve preparation of at least 4·75 litres of quite hot water, mostly for drinks. So there has to be, even given 10 min purification tablets, a supply of water anyway, with time and peace to warm it up.

Astringents, such as citric acid, understandably enough, or sorbefacients for that matter, are not prevalent among the included confectionery. Mucous and salivary mouth glands spread the flavours to the taste centres readily enough. The alimentary process in total will however tolerate wide-ranging moisture contents at ingestion—whether inherent, integral or sequence batched!

Freeze-dried foods, including rice on its own for combination with simultaneous packs of meat stews and so on, are now contrived in shapes and sizes and amounts which may have a recommendation for *boiling* water, but limit rehydration time to 5 minutes, and in many cases to much less pre-occupation, as witness coffees and teas. Ritual or relaxation may confine the attribute of speed to specific occasion, and no harm done—defusion by diffusion, as it were.

Before prolonged space flights left the realms of science fiction, imagined diets for astronauts took some whimsical forms. Weightlessness coupled with unrestricted solar light and heat prompted thoughts on 'grow one's own algae'† and 'breed lean mice'‡ for rapid

† *Chlorella pyrenoidosa* is especially high in protein and fat content.

‡ *Apodemus sylvaticus* ($3\frac{1}{2}$ inch body length) or *flavicollis* (4 in) produces 5 litters a year, each of 6 young; both were cultivated for food by the Roman conquerors in Britain.

reproduction if not satisfaction of organoleptic appetite. Large space-stations capable of containing slower food-growing systems are nearing reality, but meantime the employment of pre-prepared foods as a source of interest and enjoyment for even the most dedicated of space travellers continues to hold sway. Considerable ingenuity goes into the planning of as varied a menu as can be envisaged in an environment where platefuls are inappropriate, crumbs are taboo, liquids adopt a minimum free surface, and taste and mouth-feel form the paramount tactile solution.

It is not surprising therefore that freeze-dried foodstuffs of small bulk and high density of nutritional substance, packaged in appropriate and neat containers, demanding little except storage space, should lead the list, especially as quite a few can be comfortably reconstituted by saliva within the mouth, encouraging full mastication. Engineering has perforce had to fasten on the problem of water conservancy with a comparatively short circuit in the captive environment. Air-conditioning, cleansing and fluid flow within the body have all to be in harmony and balance as a life-supporting order wherever man finds himself, and a space-ship although cramped is no exception.

The menu selection for a cosmonaut now extends on board to nearly a hundred items, with names that are familiar to his earthly existence although perhaps not quite his normal routine. As the housewife used to be invited to participate in the minimal preparation of a convenience food by 'adding her own egg' and therefore inducing a feeling of accomplishment and pride, so certain items are arranged to require particular procedures which emulate the ritual of more mundane meal-times, and there is little doubt that this adds to personal satisfaction, easing away any psychological concern of a claustrophobic nature and holding to the adventure concept within which a picnic approach is valid.

Given such a wide-ranging selection taking up so little room, the arrival of a replenishing supply-ship may well be thought of more as a means of returning completed experimental works to Earth than as the routine grocery van. Meanwhile the mice can grow robust lapping up the algae!

A nutritional diary of a cosmonaut was reported by V. Lebedev (1985) in the Russian magazine *Science and Life*. Two-thirds of the ration on Salut-7 was dehydrated, not counting beverages or pastes and spreads, sweet/savoury, in aluminium tubes. The average daily

TABLE 6
Daily Food Ration of a Cosmonaut on Board Salut-7

Name of product	*Mass (g) net*	*Composition (g)*				*Energy value (kcal)*
		Water	*Protein*	*Fat*	*Carbohydrate*	
1st Breakfast						
1. Fried bacon	100	61·8	17·6	13·6	4·1	211
2. Potato purée	50	2·5	4·0	9·2	30·1	212
3. Borodin bread	45	18·9	2·9	1·0	21·0	99
4. Fruit dessert, plums, cherries	50	12·5	1·6	—	32·6	128
5. Coffee with sugar	24	—	—	—	20·0	75
Total	269	95·7	26·1	23·8	108·4	725
2nd Breakfast						
1. Savoury pike	100	70·0	18·0	7·0	1·3	140
2. Artichokes	25	2·4	2·4	2·6	17·6	98
3. Peach–blackcurrant juice with pulp and glucose	45	1·3	—	—	37·8	142
Total	170	63·7	20·4	9·6	56·5	380

Lunch						
1. Quail paté	100	59·8	17·3	20·4	0·4	254
2. Bortsch with meat	30	1·5	6·0	4·5	14·0	117
3. Beef tallin with potato purée	52·5	2·6	18·4	7·9	16·7	207
4. Bread, rye, Moscow	45	18·9	2·9	1·1	20·7	99
5. Sticks from quince	50	12·5	0·5	—	34·2	130
6. Milk, cow's, pasteurised	25	1·0	6·4	6·3	9·9	119
Total	302·5	96·3	51·1	40·2	95·9	926
Supper						
1. Various meats	100	60·8	17·8	10·6	8·8	200
2. Cottage cheese with blackcurrant	165	85·1	13·5	22·6	40·6	410
3. Wheat bread	30	9·6	2·7	2·3	15·0	87
4. Tea with sugar	46	—	—	—	40·0	150
Total	341	155·5	34·0	35·5	104·4	847
1. Selected caviar	82·5	52·0	3·3	11·1	11·5	157
Grand total	1165	473·2	135·3	120·2	276·7	3035

Preserves/sauce (tomato, vegetable 'Moldova', apple, cranberry)
Before 1st breakfast and lunch every day take one multivitamin (sugar-coated pill) 'Aerovit'.

consumption was 1165 g, of which 143·2 g was scheduled to be protein, 123·5 g fat, 391·3 g carbohydrates, 469·9 g the mean amount of water.

The typical menu for four meals per 'day' drawn from 83 items is shown in Table 6. Amongst other freeze-dried items reported are beetroot soup, pork cutlet, pigeon, highly seasoned stew, gruel, cabbage, cottage cheese with fruit or nuts, yoghurt, and various fruit juice concentrates in their pulp. Food containers ranged from polyethylene through cans and aluminium tubes to metallised packs. Two water 'support' systems (SRVK and Rodnik) supplied 1·7 litres per day per person, of which 900 ml was between 58 and 85°C and 400 ml

TABLE 7
Apollo XII/Lunar Module-6 Menu

Command ship		Lunar ship	
Day 1, Meal C		*Day 1, Meal C*	
Cream of chicken soup	RSB	Cream of chicken soup	RSB
Ham salad, bread	WP	Ham salad, bread	WP
Jellied candy	IMB	Jellied candy	IMB
Apricots	IMB	Chocolate pudding	RSB
Grapefruit drink	R	Grapefruit drink	R
Pineapple–grapefruit drink	R	Pineapple–grapefruit drink	R
Day 2, Meal A		*Day 2, Meal A*	
Peaches	IMB	Peaches	R
Scrambled eggs	RSB	Cornflakes	R
Bacon squares (8)	IMB	Canadian bacon, apple sauce	RSB
Cocoa	R	Cocoa	R
Orange drink	R	Orange drink	R
Day 2, Meal B		*Day 2, Meal B*	
Beef and gravy	WP	Beef and gravy	WP
Pears	IMB	Butterscotch pudding	RSB
Butterscotch pudding	RSB	Pineapple–grapefruit drink	R
Pineapple–grapefruit drink	R	Grapefruit drink	R
Grape drink	R		
Day 2, Meal C		*Day 2, Meal C*	
Turkey and gravy	WP	Turkey and gravy	WP
Chicken stew	RSB	Chicken stew	RSB
Apricots	IMB	Fruit cocktail	R
Jellied candy	IMB	Jellied candy	IMB
Orange–grapefruit drink	R	Orange–grapefruit drink	R

IMB, intermediate moisture bite; R, rehydratable; RSB, rehydratable spoon-bowl; WP, wet pack.

between 10 and 25°C. The layout of the food storage system was reminiscent of a buffet!

Commencing 8 February 1984, Soviet astronauts Atkov, Kizim and Solovyev spent 236 days 22 h 50 min on board Soyuz-T10, surely the longest sustained freeze-dried dietary experiment. By 1986, however, in preparation for a trip to Mars, at least one such explorer had spent a whole year in weightless flight, presumably on similar rations.

The American attention to the psychological as well as the nutritional aspects of space-feeding, while the journeys so far have not been so long, has been just as detailed. While a wide selection of freeze-dried items has accompanied astronauts in the last 20 years, worth mentioning especially are one-bite mini-sandwiches (thus avoiding crumbs) and 'eat with a spoon' mixtures. The latter are prone to considerable surface tension in zero gravity conditions, adhering to the back as well as the bowl of a spoon, so the feeder for (say) spaghetti with meat sauce has to have a means of scraping excess food from the back of the spoon into the pouch again. Intermediate moisture foods have been used where a base of freeze-dried product has been combined with more than one other ingredient. The research and step-by-step development was admirably reported by Klicka at a symposium in 1969. Sample menus for space flights at that time are indicated in Table 7.

Since these times it must be said that co-operation between the nations with manned space programmes has led not only to joint participation in capsule rendezvous plans but also consideration of weightlessness in the diet!

THE STATE OF DRYNESS

A dried foodstuff, in equilibrium with the atmosphere surrounding it, will emit just as many 'moisture' molecules as it receives, by definition. Being of a complicated chemical nature ordinarily, its 'moisture' molecules may well have come (been frozen out) from an intricate collection of solutions, whose concentration of companion molecules poses a certain difficulty to the 'pure water' molecules in freeing themselves to join the gaseous molecules outside. The partial pressure p that they are then able to exert, compared with the partial pressure p_0 that they *could* exert if their movement were unencumbered, is a measure of what may be termed the water activity a_w (usually expressed as a fraction) in the foodstuff under the prevailing set of conditions. If it were a case of pure water (or ice) it would proceed to evaporate at the prevalent temperature until its local atmosphere was saturated, at which partial pressure it would again be in equilibrium, for the same reason, at p_0.

The humidity of the atmosphere in harmony with the dried foodstuff, compared relatively with what it would have been were it merely embracing pure water, is still p/p_0 (usually expressed as a percentage) and therefore the other side of the same phenomenon.

Should the temperature alter, say upwards, the foodstuff will tend to balance matters by losing moisture to the air, and vice versa, until a new state of equilibrium is reached. Likewise it follows that, at a given temperature, a particular level of moisture in the reacting foodstuff, the 'equilibrium moisture content' (EMC, %) will correspond to a particular state of water activity a_w in that foodstuff, matching the instantaneous 'equilibrium relative humidity' (ERH, %) in the (adjacent) atmosphere.

Obviously a foodstuff preserved by drying, so that its homogeneous moisture content is at such a low level that virtually all deteriorative processes have been brought to a standstill, should not have that good work undermined by encountering, let alone being stored in, an adverse environment. If it does happen, the rate(s) at which exposure to such available moisture will promote take-up will depend especially on the difference of the partial pressures p (but *not* arithmetically) and will vary with the composition, both physical and chemical, of the material in its particular state of 'dryness'.

It is possible therefore to compile, albeit laboriously and with reservations, sets of the appropriate data in the form of a plane figure graph of EMC (%) against ERH (%) *at a given temperature,* termed a sorption isotherm (Fig. 20).

The sorption isotherm diagram does not define clearly a precise physical property, for several reasons, one being that foodstuff samples are, in their very derivation, rarely homogeneous in their moisture-bearing capacity; from which stems the assumption that the whole moisture content responds incrementally and *repeatably* to changes in the environment. There are in reality historical effects. The method of drying to arrive at the range in which the sample is examined, preliminary treatments, lipid amounts and distribution, the ability of the modified moisture content to effect chemical changes otherwise previously avoided, all have some sway in the result. Hence

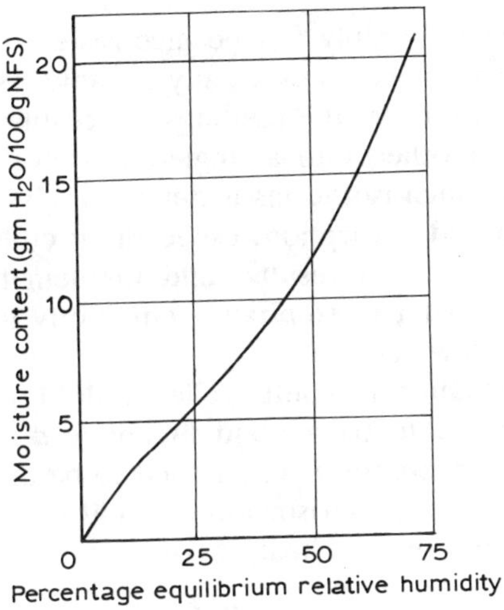

FIG. 20. Sorption isotherm for dehydrated pork at 37°C.

the background to production of the graph has to be given for guidance in extrapolation.

Preferably the moisture content is estimated on a *dry* basis, i.e.

$$\text{MC}\,(\%) = \frac{\text{Mass of moisture capable of being driven off reversibly}}{\text{Mass of bone-dry sample}} \times 100$$

The mass of the moistureless sample has to be determined by some accepted standard but arbitrary method, since decision as to actual value depends on reaching a 'no further weight loss' position, itself with inherent assumptions. This usually calls for a sample of a given size, 'spread'(!) over a given surface, to be dried to a constant weight in a vacuum oven, usually at 70°C, a prescription which requires a drying time of 6 hours or so. A variety of other methods have evolved to reduce the time required, so that the tests may be more easily utilised for quality control under commercial conditions.

The relationship at various moisture contents, possibly by weighing *in situ*, is found by exposing the sample to a series of controlled environments, usually in the closed head-space above particular solutions whose water vapour pressure in equilibrium at the strength is known. The amount of liquid, its surface area and the head-space relative to the sample size should be such that the necessary transfer (of moisture) will not alter the condition to any appreciable extent.

The solution should preferably not be inconsistent with proximity to food; glycerine, for instance, is especially suitable, as slight variations in temperature will not upset the readings noticeably.

The results are smoothed into a single-line curve for the temperature, and a family of such isotherms is obtained. Hysteresis is evident, between adsorption and desorption, especially if cycled, but a characteristic sigmoid shape often results, and mathematical analysis has been used to develop (semi-empirical) equations to fit the derived curve, mostly at the low end.

The popular Brunauer–Emmett–Teller (BET) equation, for instance, is admitted not to be a good fit above $a_w = 0{\cdot}4$, where the notion of capillary condensation is reckoned to be more apt, or below $a_w = 0{\cdot}1$, where the heat of (ad)sorption is held to be constant, but is based rather on an appropriate heat of condensation, H, thus

$$M = M_m \frac{a_w C}{1 + a_w(C-2) - a_w^2(C-1)}$$

where M is the moisture content corresponding to a_w on a dry basis, M_m is the moisture content of a monomolecular layer, $C = b\exp(H/RT)$, b is a constant approaching unity, R is the gas constant, T is the temperature involved.

These assumed relationships are used to infer a specific sorption surface area but, whereas first b and then M_m can be worked out from curve data as measured, extended tidying towards an orderly concept tends to gloss over factors of valid vacuity. This is especially true of freeze-dried material. In this case there has been a deliberate move to segregate H_2O molecules from non-uniform involvement, both physically and chemically, with other molecules, and locate them, again non-uniformly, prior to removal individually as it were in the gaseous form. Re-sorption, on the other hand, is by liquid H_2O, i.e. having a closer relationship with each other, and not necessarily attaching themselves to the deprived parts of the foodstuff immediately in the original way. Hence it is not unreasonable to surmise that (a) there was not a universal monomolecular attachment to start with, (b) some hysteresis will be due to variations in the area(s) of attachment or affinity, (c) internal energy will have some say in penetrative power, and so on, to be contrasted with what pertains in air- or vacuum-drying from the liquid phase, when translational activity in the liquid phase may be thought to provide migratory pathways for de- or ad-sorption. Whichever drying method was initially incurred, the isotherm is unlikely to have been derived during the initial desorption; repeated dryings and wettings will produce an historical element, and the *use* to be made of inferences from smoothings of these histories will be on similar but not identical histories (e.g. lipid movement). So, although M_m may well be looked on as a fraction arithmetically of M, that is all it is!

Many other formulae have been derived mathematically rather than thermodynamically, empirically to fit particular curves, or parts of curves, obtained during studies of classes of foodstuffs. Their merits and suitabilities form an interesting study in themselves. Some argument concerning the assumptions from which the various equations stem is justified, and anxiety to arrive at mathematical models of isotherms which can be incorporated in computer techniques is understandable. No one formula gives accurate results *throughout* the whole range of water activity for more than one foodstuff, and an atlas of isotherms tends to caution users into making reference only to data that most closely apply to the one foodstuff history under consideration, and the particular purpose in hand.

All that being said, without the guidance which such an idealised curve gives, the prediction of drying time to a pre-set limit, of shelf life of a dried product in a semi-permeable packaging material, and of equilibrium conditions to be reached when products of different water activities are mixed, would be all the more uncertain; from this pragmatic stance, practical workaday results do obtain.

Reduced water activity arrests the forces of natural decay in many ways. Micro-organisms are successively inhibited from growth: most bacteria by $a_w = 0{\cdot}91$, most yeasts by 0·88, most moulds by 0·80. Some (halophilic) bacteria will persist to about 0·75, some xerophilic fungi to 0·65, and some osmophilic yeasts to 0·60 or so. However, by 0·70 at room temperature, as is commonly observable, foodstuffs in general are safe from such hazards, at least for short periods. Referring to an isotherm atlas, this indicates that the *evenly distributed* moisture content will probably be well below 20% on a *dry* basis.

Depending on the drying method, other changes will occur, such as shrinkage during air or vacuum drying, and surface texture modification sometimes is witnessed during evaporative freezing prior to freeze-drying. Chemical interactions, such as enzymic activity, persist albeit slowly at lower water activities, even in some cases when pre-processing (scalding) heat treatment has been carried out purposefully to inactivate them. Lipid oxidation also can be seen to revive rapidly at low moisture levels, perhaps because the way has been opened up for it, and reactions between carbonyl groups and protein compounds, the so-called non-enzymic browning, can yet develop over years despite avoiding action in the (not so) shorter term.

In all these interactions, and in the stabilisation of aromatics, the presence of 'water' as a solvent or other constituent of inhibiting substances has a complex effect (Fig. 21) and, especially if lipids are there, straightforward removal of moisture content to the bone-dry condition is not always the best procedure, commercially or otherwise. A level of between 5% and 1% suffices for most purposes.

The intermediate range of a_w between 0·7 and 0·1 is of practical interest. In a foodstuff such as cabbage, where the original moisture content amounts to (say) 91% of the fresh weight, drying to an a_w of 0·7 equilibrates the *dry basis* moisture content at 20%, meaning that only two-hundredths of the original moisture is left. Drying to an a_w of 0·1 equilibrates the (dry basis) moisture content at 3%, meaning that only 0·3 hundredths is left.

With cooked meat, however, the moisture content prior to drying may be only three-tenths of the original (on a fat-free basis), so when drying reaches an a_w of 0·7 the EMC is 15% on a fat-free, dry basis, meaning that only some 6·4 hundredths of the *as-cooked* moisture remains, or only 1·9 hundredths of the fresh-weight moisture. When $a_w = 0{\cdot}1$ is reached, the EMC is 3%, so only just over one-hundredth

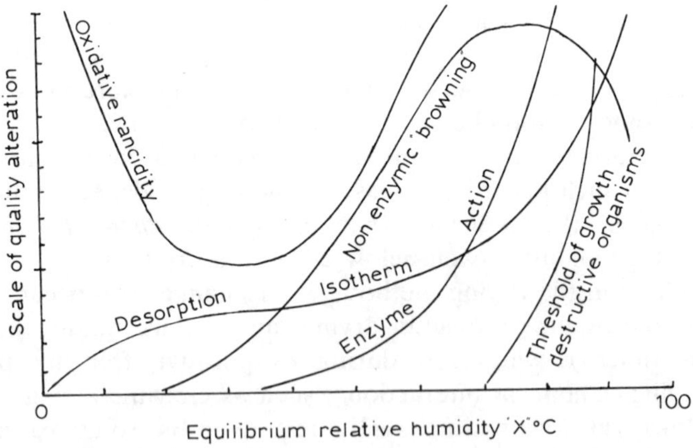

FIG. 21. Stylised relationship of factors affecting keeping qualities. Note the increased susceptibility to oxidation at very low moisture content levels (Labuza, 1971).

of the as-cooked moisture remains, or 0·39 hundredths of the fresh moisture. This is no longer protective against oxidative rancidity. Quite apart from the obvious help which cooking affords the dehydration, the general aim can be seen to be a hundred-fold reduction in the moisture content.

The 'cooking' action, as an extreme example, brought the total weight down by twice as much as the subsequent drying did, with considerably more alteration and upset of physical and chemical structure. Freeze-drying is stressed as a means of preserving the structure, or at least allowing the act of reconstitution to return the dried material virtually to the original condition, and in many (but not all) instances this is so. There restoration is not perfect; the reconstituted weight is not quite as much. Although all the other attributes come alive again, a query must be posed as to what part of the treatment reduced the (water)-holding capacity. In cooking the evidence is clear enough—cell walls are weakened, fibres are shrivelled, liquids are expressed—and in forms of drying other than freeze-drying similar modification is apparent, but freeze-drying avers that any such changes must arise in pretreatment, if at all.

The hysteresis generally to be observed under rigorous procedure, in some cases where thermal pretreatment is absent, and mostly at a_w between 0·7 and 0·1, i.e. up to 20% or so moisture content on a *dry* basis, is *not* enough in itself, although the effect may be progressive on repetition, to account for the entire discrepancy in pick-up, despite discrete pieces or substances containing little or no fat, or finishing temperatures of modest proportion at the longest-dry surfaces.

So one is tempted to conjecture that it is in the freezing part, the sacrosanct segregation part of the composite process, that something not quite reversible is inadvertently allowed to occur.

Parenchyma cells, which are most common in foods based on fruits and vegetables, consist of thin 'walls' of cellulose fibrils permeable to many types and sizes of molecules, connected somewhat loosely by pectic substances, often with air spaces between. The protoplasm within is anything but uniform, containing many different molecules apart from the essential nucleus, either in a viscous fluid or a gel.

The vacuole sections, stranded with cytoplasm, contain salts and sugars dissolved in water, and are a principal source of ice crystals when freezing is instituted. These cell sap droplets tend to coalesce on heating (as well as with age), and pretreatments that render the intervening walls weaker enable some to merge into the intercellular spaces. But the gel parts also contain water as a solvent, which may be adsorbed in trapping layers, and more difficult to free after freezing. Also the solutes may connect in its absence (owing to the subsequent drying), in such a way as to make its unstirred return much more difficult.

Animal (muscle) cells, on the other hand, although with sarcolemma friskily active in function, are very dead when submitted as foodstuffs, and usually have a history of being cooked, salted, pickled or otherwise disturbed in some preliminary way. They also tend to be naturally elongated, and therefore more likely to be dissected into several random segments.

The phenomenon of 'drip' has been pictured as the release of moisture previously expressed irreversibly from proteinaceous elements during lowering of temperature, at a *slow* rate, to far below the accepted freezing level(s), resulting in the massive growth of crystals, and is now knowingly avoided.

Although there is, by further contrast, likely to be fat rather than air interwoven with the connective tissue, leading to stretches of 'wall' where moisture may be repelled, especially after cooking, the complexities from within these cells—and their variety is much wider ranging than in the living vegetables—still promulgate irreversibility puzzles in the literature.

None of these (perhaps overstressed) troubles seems to be manifested in 'hospital' or pharmaceutical freeze-drying operations, where temperature plunges, stylised quenching and controlled tempering are the rule.

Much data in the literature on foodstuff freeze-drying have less intensively annotated background descriptions of preliminary treatment; therefore one has to conclude that at least some of the reported inability to reconstitute fully may be due to commercial adaptation of the freezing operation. Incidence appears to be lessening and, in any case, since no industrial outcry or consumer anguish has been voiced, the phenomenon may have been less widespread than the academic reportage indicated.

Returning then to the present discussion of isotherm hysteresis, which does seem to include some freeze-drying histories, there are still two points to be made. Transfers of atmospheric vapour to and from an already processed substance, as may be encountered in responsible storage, are unlikely to entail the same extremes of sensation as have had to be occasioned in the derivation of desorption and adsorption data by the means outlined above, which themselves may have caused the alterations in the specimen exhibiting the hysteresis.

Secondly, the rate of change of moisture experienced in post-drying encounters, say with partial pressure(s) of vapour in head-spaces, such as of coffee granule jars or powdered soup sachets, or in blendings and mixtures of larger particulates, where the sheer quantity itself protects the exposed/unexposed surface, is relatively slow and is fairly readily forecast mathematically.

Since it is somewhat impracticable on full-scale plant to determine the arbitrarily chosen 'end' of drying by interruptive weighing to find out if the average moisture content has reached an acceptable level, the a_w level, or rather the partial pressure of water vapour above the nearly dry product at the estimated temperature, can be measured. This, as a quotient with the saturation pressure, from tables or Clausius–Clapeyron equation(s), and referred to a tuned or standardised isotherm (formula) for the substance, will allow practical inference of the mean moisture content, and hence a decision (by the computer) as to when the drying process should be curtailed. The latitudes are allowable.

Note that, in arriving at the equilibrium condition in a test (Tables 8 and 9) humidity over a known solution, considerable time (days) was allowed to elapse, and care taken to maintain a close watch on the temperature. Here, during a drying run, the temperature within the virtually dry piece is unlikely to be so stable, or correspond to a static reference partial pressure except briefly while isolated, which is more difficult towards the end of a run than near the beginning. Such reservations have to be accommodated, and a factor is sometimes

TABLE 8
Relative Humidity over Glycerine Solutions

Refractive index at 25°C	*Relative humidity (%)*		
	0°*C*	25°*C*	50°*C*
1·3773	89·2	90·0	90·7
1·4015	78·8	80·0	81·1
1·4191	68·6	70·0	71·3
1·4329	58·4	60·0	61·4
1·4440	48·3	50·0	51·5
1·4529	38·3	40·0	41·6

TABLE 9
Relative Humidity over Sulphuric Acid Solutions

Density at 25°C (g/cm^3)	*Relative humidity (%)*		
	0°C	*25°C*	*50°C*
1·375	36·1	40·0	43·0
1·425	26·5	30·0	33·0
1·478	17·0	20·0	22·8
1·545	8·1	10·0	12·3
1·610	3·7	5·0	6·3
1·7175	0·5	1·0	1·3

employed to forecast the likelihood of the next reading. This factor is the inverse of the slope of the isotherm graph at the point. Away from actual drying, where the dry tissue may be subject to steep gradients of temperature with thickness, determinations affecting storage equilibria are much more straightforward.

Over a thousand reference graphs have been assembled in a handbook of food isotherms (Inglesias & Chirife, 1982). A quarter of them mention freeze-drying, either as the original method of dehydration or as part of the preparation of a sample dried initially by other means. The 130-substance collection is unable to portray consistent process or treatment differences. Use of EMC/ERH patterns like these depends on the *temperature* relationship, so isotherms plotted at limiting upper and lower levels are preferably to be sought. Within these, curves at discrete intervals have to be mapped. Extrapolation requires caution, since calculation is by no means clear-cut. It is worthwhile to take pains in assembling significant operational data of this nature, so that the extent of its reliability and therefore the extent of hypothetical risk may accurately be judged.

CODA

Whilst foodstuffs usually comprise (recently) dead organic matter, and therefore do not pose all the problems associated with the preservation of living things in a dried state, most of the orthodox techniques have been established from studies on bacteria, viruses, yeasts and the like, which can survive freeze-drying, having learned by evolution in the dry cold of Space.

Many proteins need not denature or aggregate irreversibly; it is often the cell wall enclosing them that is in danger of being ruptured during the freezing stage. Blood plasma, the yellowish *carrier* liquid, has been freeze-dried abundantly, but not whole blood, because of this.

Glycerine, which helps certain insects to survive freezing in Nature, has been the translation means of protecting live sperm cells and (later) red blood cells from descent to −80°C and storage there. Return ready for use involves sterile washings to purge the glycerol cryoprotectant. However, successful drying (and therefore freedom from the means of a continued frozen state) of such delicate cells is not so clearly delineated. Present circumventing practice leans to taking the oxygen-transporting haemoglobin molecules *from* the (broken) blood cell membranes (of outdated stock), treating them chemically to prevent their tendency to break up, and freeze-drying them *naked* as it were. Rehydration by saline solution is then intended to provide a blood substitute conveniently in emergency situations, especially in places where frozen supplies would be a problem.

The problems for foodstuffs are nonetheless complex. Reconstitution can, for the most part, be assessed by texture, or rather a lack of resultant flaccidity. Excipients are not in favour except as inactive carriers afterwards of additives that have to be declared and explained. Fortunately such posers as oxygen-carrying capacity do not apply, nor is cell-wall binding of proteins an essential requirement for the intended purpose of a foodstuff. Organoleptic criteria are paramount, in contrast, since nutritive assimilation has otherwise been assured.

Freeze-drying of micro-organisms for the food industry is formidable business. Yeasts, especially chosen for baker and brewer, are budded and dried in vast amounts for consumer product use. They are cultivated for their by-products. It may be going too far to say that no one, apart perhaps from live yoghurt enthusiasts, conspicuously advocates bacterial ingestion to aid assimilation of foodstuffs in the human gut.

Yet the manifold activity of bacteria is vitally important in the food chain, earlier in the nurture of plants for instance, and terminally in the scientific disposal of waste. Current work of interest, on crop protection, includes the fashionable genetic engineering of microscopic organisms (i.e. employment of sexual mechanisms other than binary fission) with a view to directing their efforts more succinctly. A major difficulty has been persuading authority that unleashing mass-produced gene-spliced 'new' derivatives mutant into the environment has no untoward side-effects. The following are examples of quite bold experiments.

Pseudomonas syringae lives on the leaves of many cultivated plants. It promotes the early formation of frost when temperature hits 0°C, because a protein on the capsule membrane acts as a nucleus for an ice crystal of dew, leading in due course to freezing of the plant cells themselves. Ingeniously, by recombinant DNA techniques, part of the gene ordering production of the ice-nucleating protein has been removed (or impaired) so that, when several trillion altered microbes are sprayed on (say) a field of strawberry plants, they can withstand unseasonal temperatures of down to −4°C for a short time, and the extended growing season is not jeopardised.

Pseudomonas fluorescens has been endowed with a gene from another bacterium which produces a protein deadly to cut-worms (soil-living caterpillars which gnaw at stems and roots). Seed-corn coated with the engineered microbe develops roots in the growing plant which contain defending colonies. Mass production of such mitosis has to be contrived, as normal conditions of growth may produce a likelihood of mutation in the manner desired, at a rate of only 1×10^{-10} per bacterium per generation, depending on the species.

Freeze-drying permits commercial quantities of such 'helpers' to be prepared and distributed. An important body of 'helpers' is the category of proteins called enzymes, which are natural catalysts widely used in food processing. Direct designing and mass-production of these cheaply is a major aim. For instance, yeasts have already been engineered to make sweeteners, to make rennin for cheese production, while oil-bearing seeds are being vectored to modify their fatty acid composition. The isolation inherent in the freeze-drying process is consequently attractive.

Chapter 6

Physical Considerations

Amadeo Avogadro (1776–1856), definer of the molecule in his hypothesis that equal volumes of all gases under the same conditions of temperature and pressure contain the same number of 'divisible parts', also assumed (1811) that these molecules were equally distributed throughout space in the case of all gases.

It assists comprehension of the physics of freeze-drying if such elementary conceptions are borne in mind, along with the idea that a *vapour* is the gaseous state of a substance *normally* liquid or solid.

Thus fog or cloud is transient, being vapour condensed about dust nuclei, the minute particles of 'water', sometimes in a frozen state, floating in the atmosphere, i.e. amongst other gases. Of all these present in the terrestrial atmosphere, H_2O is the only one that can exist comfortably in any of the three states of matter (Fig. 22), and its condition is the major influence on total barometric pressure. Because of its temperamental changes of state, it is the partial pressure of water vapour in the atmosphere that fluctuates enormously, the other partial pressures nearly all remaining virtually static, although in motion as components. Some (man-made) products, generally of combustion, the natural cycle of respiration, and the penetration of heat, account for most of the minor fluctuations.

Normally in contrived freeze-drying we deal with captive environments where the 'atmosphere' is controllable and totally known, but in the open air the collection or dispersion of the H_2O molecules depends on the random appearance of its local *partial* pressure and the local general temperature taken together.

Incidentally, 'steam' is often used as a transitional term for liquid droplets and vapour existing simultaneously in the same place, when

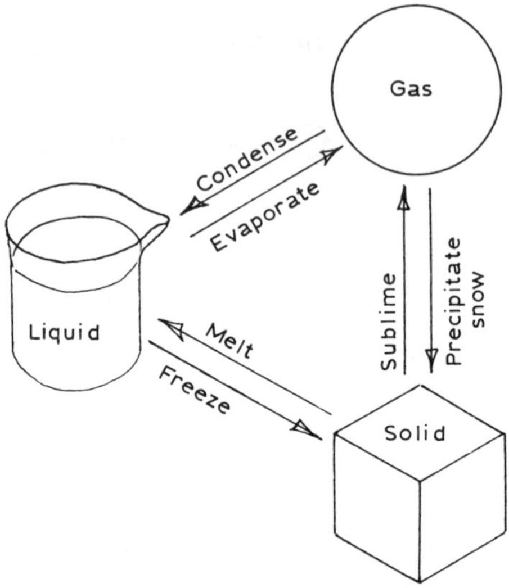

FIG. 22. Transformations between the three states require known exchanges of energy, whichever route is taken.

supply or denial of heat energy will change its state at constant temperature. Vapour superheated above the saturation temperature for its pressure is a dry, invisible, transparent, rule-obeying gas.

The vital clue, given by Avogadro so long ago, and not then really well received by chemists, is that there is enough room between freed molecules for them to be treated as existing as if there were no other families of molecules present and, if chemically inactive, no attraction to them either. But they do respond to the general activity or temperature encounters around them. If there is ample space for the H_2O gas to fill, or a 'sink' continuously removing it, or the temperature changes, then the pressure (retro-propulsive action against surroundings) *belonging to the H_2O vapour itself* will be affected.

The temperature is of course a measure of the potential activity of molecules, confinement leads to pressure, and the two are complementary. Entropy, rather than Nature, abhors a vacuum.

In contrived freeze-drying therefore, performed in a finite space, we supply heat energy to (H_2O) molecules in the solid state *at the equivalent rate* as we remove molecules in the gaseous state, either numerically from the chamber *or* by abstracting energy to change their state, numerically from the 'free' space. Although the intention may be preferential, heat will be accepted by other matter in the vicinity, which will tend to pass it on imperceptibly. Reaction to variation of the energy supply to the process itself is virtually immediate.

A great deal of study has centred on endeavours to improve the gradients along the difference in vapour pressure between the 'source' and the 'sink', on which depends the driving force for sublimation. The ice-front temperature determines the vapour pressure at source, and resistance to escape from the preliminary labyrinth depends on how far that front has receded from the surface of the piece. So the 'enticement' of the sink has to cater especially for the end condition.

The sink may be the throat of a steam ejector, the surface of an 'ice condenser', or some other device that contrives a lower vapour pressure to which the expanded vapour is attracted. Collapse in volume of the vapour represents an enormous pumping action, and may be occasioned by change of phase from (superheated) gas to snow/frost/ice deposit, *or* by chemical adsorption.

This latter method is currently being investigated by John D. Mellor (1988, pers. comm.; Bell & Mellor, 1986) using silica gel (graded 4 mm) surrounding museum specimens and wild flowers, or arranged without actual contact if foodstuffs like abalone, parsley, oregano and the like are being dried. Retention of distinctive aromatics is particularly stressed. To achieve similar low vapour pressure, a refrigerant would have to be operated at very low temperature, with consequent reduced capacity and higher cost (see Table 10).

Abalone 5 mm thick has been dried to good quality in under 24 hours in finned trays at 0–5°C while the cabinet and other contents,

TABLE 10
Comparison of Vapour Pressures (mm Hg)

Temp. (°C)	*Vap. press. above ice*	*Vap. press. above silica gel*	*Moisture content of gel (% dry wt)*
−20	0·776	0·0026	9
		0·026	15
		0·18	22·5
		Change gel and regenerate	
−40	0·096		
−50	0·03		
−60	0·008		
−90·57	0·0026		

including the gel, were all at −20°C. This chilled desiccant system prior to final pack avoids 'in-can' desiccation.

Lowly Egyptians, in the pre-Dynastic period (i.e. more than 5000 years ago) were buried rather simply in the hot dry absorbent sand at the edge of cultivation. There the body dried out, to a baked skin shrivelled over a grim skeleton, surrounded imaginatively by storage pots for food and drink on the long journey into the further world. Later methods of 'embalming' dwelt on preserving the superficial appearance of the eviscerated corpse, so the main dehydrating agent became natural compound crystals from Wadi Natrûn, near Cairo. Sodium carbonate and bicarbonate, with admixtures of sodium sulphate and chloride, has an affinity for absorbing water and is mildly antiseptic; in some 40 days of a formal total of 70, dry-salting of human remains was complete, ready for the wrapping with linen bandages. Much less attention seems to have been paid to the preservation of actual food accompanying the ritual entombing of dignitaries, protocol relying instead on wall painting instructions to the spirit forms on how to set about growing, producing and preparing food offerings. By the 21st Dynasty, as hedged bets, Tutankhamun had *with* him fruit, wine, roast duck, loaves, cakes—and *sun-dried fish*!

Long beforehand, New Stone Age man was preserving grain in moisture-drawing straw-lined pits. The Incas, quite separately, learned to desiccate in rarified cold mountain air. It must be concluded that, although dehydration in general became widely enough known and practised, the means of so doing had stylised boundaries.

Rational pressure and thermometric scales are, of course, relatively modern concepts, as are the sophisticated procedures of process engineering. Freeze-drying had to wait until this century to be thought out deliberately as capable of being constructed, tailored to the task, and organised in a business framework. World-wide, however, commercial freeze-drying of foodstuffs is currently achieved by contriving a vapour pressure gradient consistent with the use of mechanical refrigeration.

FREEZING

When a pure element solidifies, by concept it does so at a fixed temperature, its freezing point. A cooling curve of temperature against time will thus show a horizontal portion at that level, due to the relinquishment of 'latent' heat. Without any suitable solid matter to cling to, crystals may have difficulty in starting to form. The liquid may supercool as much as 3° below its real freezing point, as it tries to make its own seed or nuclei, from which crystallisation can (but need not) suddenly set in. The temperature, unless heat is being extracted very rapidly, will then rise to the true (!) aforementioned freezing point.

Attainment of a non-crystalline, amorphous, vitrescent, solid state is quite possible in such a metastable condition, but if the labile (unsteady) region *can* be entered, when many tiny crystals form, these will fuse into larger ones whose shape will be modified by the history of the circumstances. Thus, while dendrites will grow from the nuclei corners until high surface tension or neighbouring growths round them off, and crystalline forces form them towards geometrically simpler shapes, viscosity and/or liquid/gas currents (depending from which phase the solid condition is coming) will also play a part, but the time allotted is of the essence.

When a simple solution has heat abstracted uniformly and steadily towards the solid state, then one part, the instigator, the solvent, will tend to solidify first, in so doing divorcing itself from the solute, which may be said to re-group to a stronger solution. From the point of view of the growing crystal, such 'impurities' will get pushed aside until they are lodged between the 'grain' boundaries, or are trapped in the arms of the dendrites before they themselves dissolve physically into the large crystal or grain. These occlusions may be the only hint of how the solid entirety was formed.

The ever-concentrating segregated solution meanwhile finds it more and more difficult to settle into a change of state, pictured by ascribing a steady lowering of the hiatus in temperature at which this would be possible, together with a snap reaction back there should it succeed. It follows that, when very much more complex solutions are cooled in like manner, the various solvents, where there are more than one, will tend to extract themselves or dissociate (if that is an acceptable word pictorially in the context) according to some recognisable order of precedence on the temperature scale. Fortunately peritectic reactions are not so demanding of our present attention as are eutectics.

The pragmatic purpose of thorough freezing preparatory to freeze-drying is specifically to segregate the H_2O molecules in solid form by themselves, and in so doing to render the remaining constituents of the biological complex, however many and however involved, temporarily inactive chemically.

(*Note*: Oliver Lyle, in his *Efficient Use of Steam* (1968), puts forward the premise that, while *chemically* water can be regarded as a hydroxide of hydrogen, H(OH) (hydrogen ions or acid producers, and hydroxyl ions or alkali producers), its *physical* behaviour suggests that it consists of a solvent di-hydrol, $(H_2O)_2$, carrying in solution at high temperatures *steam* molecules, hydrol (H_2O), and at low temperatures *ice* molecules, tri-hydrol $(H_2O)_3$. This theory explains changes in specific heat, density and so on. It does not mention deuterium, the isotope of hydrogen with an atomic weight of 2, or tritium, but perhaps for the understanding of this process our interpretation of 'water' in and/or out of foodstuffs is heavy (!) enough without that.)

In the knowledge that such systems inherently object to remaining inert, the key is doubly turned in the metaphysical lock by removing preferentially the ice as its own pure vapour, or seeking to do so, with the intended effect that minute and exceedingly slow readjustments during subsequent isolated storage at more normal temperatures simply cannot occur. For this to be achieved, not only must the subliming vapour be able to find its way out of whatever structure contains the ice, but the contrivance of the act of solidifying by an arranged form of cooling should, if possible, aid and abet its favourable positioning. Also, the initial stage of separation must pay heed to the condition or state of what is left behind and how indeed it reached that state—a not dissimilar aim to Nature's protective precautions against severely wintry weather.

Ideally, culinary intention of itself should dictate the general dimensions of the preserved foodstuff piece or dried beverage particle, but these sizes may well be strongly influenced by commercial processing considerations. Temperature gradients within the piece as it freezes will reflect its physical peculiarities, changing concentrations, shifting thermal conductivities. How it freezes will determine how it dries. Preparatory steps ensure similarity but not sameness.

Cut and surface cells will tend to develop slightly different, elongated crystals compared with those near the heart of a piece (Fig. 23). Scalded vegetables will have lost their intercellular gases, filling the spaces with liquid which, depending on the method, will have come in part through the modified cell walls. Appropriate to the level of dissection, whether flesh of fish, fowl or higher animal, the relative position to the cut edges, of muscular strata, connective tissue and fat,

FIG. 23. Evaporative freezing of wet-surface slabs of fish was effectively carried out rapidly by pulling the vacuum in the chamber. At Aberdeen use was made of perforated plastic sheet between product and expanded metal insert (a) to avoid heat input from the metal shelf and (b) to eliminate the 'case-hardening' (formation of a pellicle or hard skin) which otherwise might form. Seventy 2 mm diameter holes/in^2 gave the 'open' area as one-third.

will in the main be random, and therefore the freezing action will advance from different external surfaces at different rates. Even thin soup purées or fruit juices which are devoid of organised form, but contain wisps of parent matter, and are frozen in such a manner that they have one thickness dimension which is uniform, may be particled so that other dimensions have latitude above and below it. All these internal aberrations, perfectly normal as regards culinary acceptability, prompt the employment of ultra-cold 'sinks' to speed the attainment of final uniformity.

A variety of chilling and freezing methods are employed for the foodstuffs, once ratio of surface to volume is established. Total immersion, popular for clinical work, is rarely used. Pre-chilling chambers, blast tunnels and fluidised beds (for 'individual quick freezing', IQF) are as likely as contact-shelf, drum or scraped-surface styles, while evaporative freezing by vacuum-drawing on batches is not uncommon.

The original recommendation for fish, 'rapid' freezing, defined as taking no more than 2 hours to pass from zero to −5°C (freezing point being reckoned as −1°C) in the *thickest* part of a sea-creature, has been extended as a rough guide to cover other products. Its origin was partly to stem the multiplication of putrefying bacteria and partly to reduce the amount and size of crystals histologically determined as growing between the cells; freezing at twice that rate was reckoned to produce small crystals mostly *within* the muscle cells. What is to be preferred more specifically might be for water to be osmotically drawn from the cell through *intact* sarcolemma to form crystals outside, which are too small to distort fibres, provided of course that the muscle proteins on some subsequent occasion will as readily take back such water. Compromise is inevitable in normal freezing practice.

Storage of fish between −23 and −30°C is now regular practice, but achievement of these lower levels is often at slower rates in cold rooms. A level of −18°C is typically aimed at for *most* produce. Meat in carcass form has to undergo conditioning of pH for a time about +10°C, then ageing around +4°C before proceeding to part butchering and/or freezing in suitable halls, followed eventually by full butchering. The phenomenon of 'drip' indicates the extent of cell-wall damage. For freeze-drying, sub-cutting of the large carcass pieces is to be preferred at an earlier stage. Truly the state of being 'very dead' makes a difference.

Before one may specify a cooling rate at which solidification best may be arranged, or where the pure, segregated ice is best disposed, one has to decide what measure or 'degree' of solidification will be sufficient adequately to lock up (i.e. freeze) those chemical actions that might adversely affect preservation of the material for the purpose envisaged. It may well be presumed that these will be sufficiently deterred if the substance is kept below a particular temperature level, although we have seen that slowly to approach lower and lower temperatures makes for continuous concentration(s) *and* an ever-receding target.

What is desired is a freezing of all the (concentrated) constituents into coexisting solid solutions, each with its own kind of crystal, as it were, to form eutectics, *which condition will be reversible* despite subsequent drying.†

† That this can be done without harm is evidenced by the current use of bovine semen that was plunged into liquid N_2 in 1963 and held for more than 25 years at −196°C.

This state can be gained by a very deep plunge of temperature, as may be achieved by spraying with, followed by immersion in, liquid nitrogen at −196°C. It will then be found, on gentle warming, that there will be reached a true level of temperature at which there is incipient melting. Freeze-drying as such must be conducted with heed to this level.

Prior to reaching this level, from such a depth, it has been propounded that, for a number of (perhaps all) substances, a level will be found at which there appears to be a devitrification or (re)crystallisation evidenced by some exothermic reaction. Above such a level, sublimation of the desired constituent may be arranged. Thus limits of permissible range of temperature of the *residual* frozen mass of material during freeze-drying have been arrived at.

The argument owes something to metallurgical science. It was first checked out as pertaining to the realm of foodstuffs by Louis Rey (1959), studying solutions similar to and including fruit juices. He employed measurement of changes in electrical resistance to detect when complexities froze or melted, based on a proposal by Greaves that electrical conductivity of interstitial fluids would be affected as they were squeezed out of connection by the growth of ice crystals.

Since then, a number of foodstuffs have been tested by apparent freezing according to conventional freezing techniques, later tempered by cooling to −196°C, and then raised through the observed levels to show repeatable results. Whether it is indeed necessary to go through quite such a rigorous procedure in practice with cellular (dead) foodstuffs is a moot commercial point, it being held in some quarters that, given the limits of level thus established, cooling down reasonably rapidly (*sic*) to within the band and drying from there often produces product that exhibits either no apparent deteriorative storage over the intended life-span or none that virtually matters. Such departures in procedure could not be tolerated in clinical freeze-drying of course, where the internal (heat) energy not given up on vitrification is probably essential to survival of the specimen, but calculated risks concerning what activity may still be possible hygienically within a preserved foodstuff, having taken all other precautions to prolong its survival in an acceptable condition, are not beyond the bounds of probability.

While there are published data on the beginning of melting for eutectics of several specified sugars with water, for instance, and for a number of fruit juices which contain similar systems as well as additional components, their very lack of insoluble structures has fastened attention on what is termed their 'collapse' level. This is reached in drying when the nebulous scatter of residuals, disconnected by the isolating ice formation, cannot hold together as a candyfloss attenuation. It seems unable to accept further heating without succumbing to a kind of puffing ebullience. The hazard is avoided by solidifying as a thin sheet with incorporated brittleness which is shattered prior to drying in a particulate bed, observing due caution throughout. Obviously pre-concentration aids all parts of this procedure.

Detection and identification of intermediate points of temperature arrest arising from changes in such conglomerations of solid solutions are difficult remote from the specially adapted stage of a microscope containing a conductivity cell. Just as hydrogen-ion concentration (pH) has been made a measure of the departure of a solution from neutrality, electrical conductivity (Rey, 1959) is used as an indication of the activity or functioning of *all or any* of the ions present in the solution or mixture of solutions. *This* reciprocal, of electrical resistance, can thus be used where a myriad of tiny pathways exist between crystals, and continues as these are disconnected or closed, even while one intricate labyrinthine pathway remains. Beyond this, which is devised using alternating current to avoid polarity, one has to consider how effective dehumidification of the tissue that is there has produced insularity, if not retention of charge which is the characteristic of organic compounds. There are many propositions on these matters. Deep-frozen electrical resistance may rise from better than 10^{12} ohms to around 10^{8} ohms before second-order phase transition then ante-melting is observed as very slight changes in slope of a warming curve progressing at (say) $\frac{1}{2}$°C per minute. Fruit juices exhibit broader bands to maximum melting than individual sugars. Raw meat data give higher values than cooked meat data, both based on minced samples. Translating these to working practice on a production scale for foodstuffs might in many situations lead to cutting the Gordian knot, i.e. reliance on embedded thermocouple inference.

Around 1954, Stone & Webster Engineering Corporation in Boston, USA, patented a 'ripple' tray to rival the traditional 'bubble-cap and sieve' system used in two-phase fluid contacting columns familiar in distillation or absorption plant towers. It not only was perforated, but had sinusoidal waves pressed in, so that the perforations dripped only at the trough, and were splashed over elsewhere in the wave.

An advance at Aberdeen, Scotland, was made on this idea to solve the steam scalder end-tower problems reported from Western Regional Research Laboratory, Albany, California, by using expanded metal arranged meander fashion, such that the wash liquid, persuaded by the parting and coming together of the mesh strands surrounding each rhomboidal aperture (Fig. 24), tended to film over it. The steam,

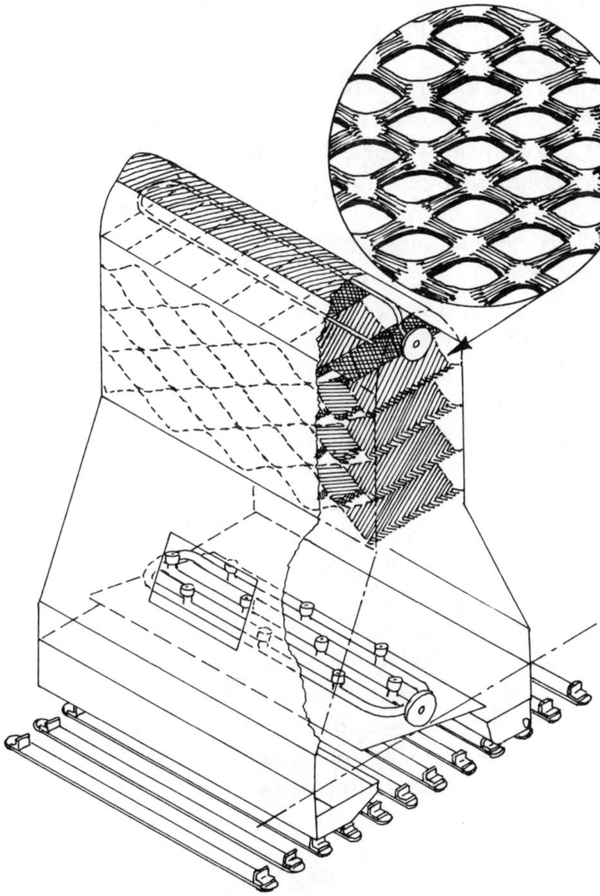

FIG. 24. End tower of scalder. The top 'hairpin' water spray wetted a series of undulating expanded metal meshes; the liquid ran down the slopes, filming over the gaps. A contra-directed series of jets induced external air and internal steam to move upwards, bursting the films. The steam was thereby *gently* condensed, mostly falling as rain to the collecting troughs. These balanced chambers effectively reduced draughts, loss of steam and end-cooling.

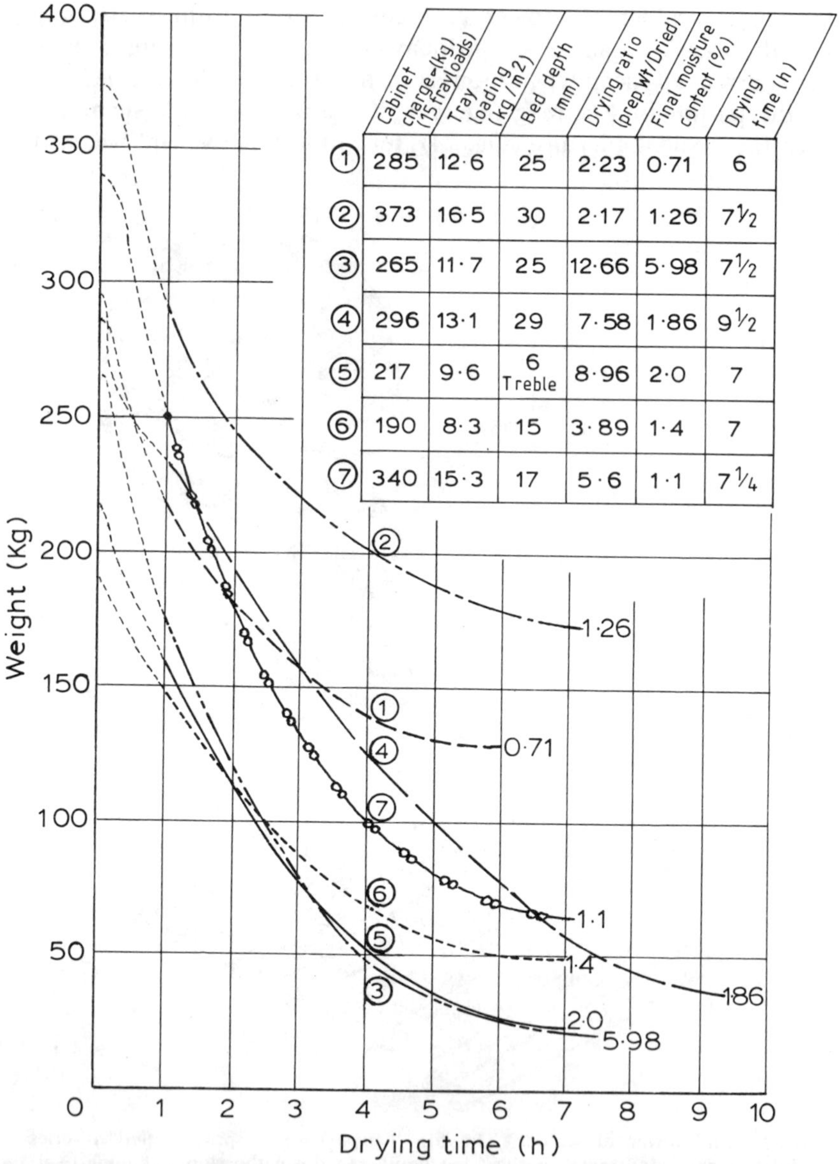

	Cabinet charge-(kg) (15 trayloads)	Tray loading (kg/m2)	Bed depth (mm)	Drying ratio (prep.Wt/Dried)	Final moisture content (%)	Drying time (h)
①	285	12·6	25	2·23	0·71	6
②	373	16·5	30	2·17	1·26	7½
③	265	11·7	25	12·66	5·98	7½
④	296	13·1	29	7·58	1·86	9½
⑤	217	9·6	6 Treble	8·96	2·0	7
⑥	190	8·3	15	3·89	1·4	7
⑦	340	15·3	17	5·6	1·1	7¼

FIG. 25. Weight loss curves and conditions for a range of AFD products on the Mark I cabinet scale. 1 and 2, cooked pork, minced; 3, scalded cabbage shreds; 4, dipped raw apple, diced; 5, ditto, rings; 6, raw beef steaks; 7, raw cod steaks. The rapid 'tail-end' peculiar to this (insert) process adjunct is especially noteworthy.

helped by the air stream, had to get through this 'bubble' cascade or perish! The scheme was effective in comparatively little height, isolating the chamber from air draughts, and condensing the wire-drawn steam from 103°C or so at atmospheric pressure, without inducing excessive steam consumption attracted to the towers. Other forms of steam scalder now, of course, exist.

It was while scrutinising the expanded metal from all angles, during the development of the scalder tower, that the idea occurred of using its sharp points against the product, its rounded side against the heating surface, as an insert between product and double-contact heating platens, so making freeze-drying possible. By this means a pattern of heat transfer would be maintained while its intricate aperture pathways (see also Fig. 35) would eliminate the stifling effect of the flat surfaces of product slab and heating plate. In the event the inserts 'accelerated' the process to commercially advantageous throughput times (Fig. 25).

Heating means are described in the next chapter.

THE PATHWAY WITHIN THE PIECE

Nature is complex, incredibly intricate, and variegated; her products for the table are no exceptions. So when attempts are made to prophesy production rate by calculation of the time to dry discrete portionings of such a substance, for instance, one has to accept the scant regard for precise homogeneity which man-made morsels, in that sense, present. It is then perfectly understandable that the device of turning the factual problem into a hypothetical, easier problem is resorted to, provided that one remembers that the purpose is not wholly to find an exact solution to the *real* course of action but, by applying fictitious constraints to the variables in turn, to appreciate their individual effect on the whole.

Unfortunately this aspect is not always borne in mind, in the natural anxiety to provide a simplified model that agrees, within an arbitrary percentage of result, with common repeated experience. Not that this should be decried; it may be the only tidy way of approaching a general uncertainty, or at least allowing one an insight commensurate with the level of understanding necessary for practical purposes.

Dividing the working day of 24 hours into one, two or three 8-hour shifts has had more effect on production schedules and product dimension and loadings than almost any other assumption. The danger, if any, of going by the model may only be in that wider application ensues which overlooks the early tenets.

For example, particularly within the ice-emptied traceries that constitute the developing pathways for the escaping vapour amid residual non-condensable or 'permanent' gases, where such may abound, the simple picture is of 'flux' being equated with a 'concentration gradient' times a (–) 'diffusion coefficient' D. This constant of proportionality then has to have units of L^2T^{-1} and the concept is a confine of a *long* imaginary tube.

Assuming either the one-quarter or the one-sixth model of a (perfect) gas, and that the free paths are all equal to λ, then $D = (1/3)\lambda\bar{u}$. Even when it is accepted that the atoms move in all directions and have a range of free paths, such that there is a probability of 0·63 that an atom will suffer a collision in travelling a distance equal to λ (Walton, 1983), this definition of D conforms. Again, the mean speed $\bar{u} = (8kT/\pi m)^{1/2}$ and $\lambda = kT/\pi d^2 p$, so

$$D = \frac{1}{p}\frac{1}{3d^2 m^{1/2}}\left(\frac{2kT}{\pi}\right)^{3/2}$$

from which $D \propto p^{-1}$ or $D \propto T^{3/2}$ but *also* $\lambda \propto T/p$.

In a Knudsen gas, the atoms travel without collision from one side of the 'tube' to the other. There has to be more gas atoms at one end of the (foodstuff) tube than the other for flow to occur, but at every impact with the 'wall' the conditions that specify D will have altered. Hotter wall(s), superheated gas, the temperature variation has more effect than the instantaneous pressure!

Another favourite concept is the introduction of a regular conductance U from $Q = U(p_1 - p_2)$ where $U = (\pi^3 kT/2m)^{1/2}(r^3/l)$ indicates that larger bores and shorter lengths of capillaries will aid flow—a common-sense deduction. How to contrive this, other than academically or in the glassware of the laboratory, is of course why interest is shown by the processor in smaller rather than larger cuts, in direct conflict with the view that one of the main features of freeze-drying foodstuffs is the ability to furnish recognisable pieces.

With persistence and the application of a number of factors and 'constants', the recalcitrant experimental results can be fitted to the model with satisfaction; this has been the focus of numerous investigations. Equally, it has required such definition of prescriptive boundaries that, as suggested earlier, pragmatic empiricism has not so far, for foodstuffs at least, been swept to one side. After all, an 'equivalent circular pore' is as pragmatic an academic device as one might wish for.

Hypotheses dealing with the happenings between ice front and the chamber interior, in some cases as far as the refrigerating vapour receiver, are exhaustively reported in the mathematical sections of Mellor's book (1978); it is, however, only fair to say that quite a number pertain to high vacuum, low levels of temperature and relatively small pathway apparatus, such as are more usual in medical freeze-drying. Queries develop when these are tested against full-scale foodstuff freeze-drying 'swirling/scudding' experience, as Mellor readily admits in comment. The underlying difficulty is the wish to provide a universal set of formulae covering material to be dried which is randomly varied in form and substance. The models then require explanation as 'chiefly mineral, with animal and/or vegetable connections'.

Strictly, the diameter of a molecule (or an atom) is a concept rather than a reality, although it might be considered to have a sphere of influence in relation to other molecules, depending on the circumstances or indeed the assumptions. Most pertinent theoretical discussions are based on one kind of molecule at a time, or consider hypothetical situations where one set of molecules (e.g. air) displaces another set of molecules (e.g. water vapour), the first set moving principally in one direction whilst the second set conveniently is stationary or moving in the opposite direction. Outside the foodstuff labyrinthian manoeuvres, however, two such sets are probably for the most part moving together, although perhaps at different temperatures, partial pressures and speeds.

TABLE 11
Molecular Diameters

	Relative molecular mass M_r ($\times 10^{-3}$ *kg*)	*Solid density* (*kg*/m^2)	*Molecular diameter* ($\times 10^{-9}$ *m*)
Air	29·1	330	0·1464
Nitrogen	28·013	311	0·1496
Helium	4·003	69·3	0·0096
Water vapour	18·016	917	0·0326

Based on $D = [M_r(X) \times 10^{-3}/\rho N_A]^{1/3}$, where ρ is the density in the solid state and N_A is the Avogadro constant.

TABLE 12
Molecular Free Paths (mm)

	Partial pressure (mm Hg)	*Air*	*Nitrogen*	*Helium*	*Water vapour*
At +60°C	0·2	2·56	2·45	59·49	5·16
	0·33	1·55	1·49	36·05	3·13
	0·66	0·78	0·74	18·03	1·56
	1·00	0·51	0·49	11·90	1·03
	10·0	0·05	0·05	1·19	0·10
	20·0	0·03	0·02	0·60	0·05
At +20°C	0·2	2·25	2·16	52·35	4·54
	0·33	1·36	1·31	31·72	2·75
	0·66	0·68	0·65	15·86	1·38
	1·00	0·45	0·43	10·47	0·91
	10·0	0·05	0·04	1·05	0·09
	20·0	0·02	0·02	0·52	0·05
At +3°C	0·2	2·12	2·03	49·31	4·28
	0·33	1·29	1·23	29·88	2·59
	0·66	0·64	0·62	14·94	1·30
	1·00	0·42	0·41	9·86	0·86
	10·0	0·04	0·04	0·99	0·09
	20·0	0·02	0·02	0·49	0·04
At −7°C	0·2	2·04	1·96	47·52	4·12
	0·33	1·24	1·19	28·80	2·50
	0·66	0·62	0·59	14·4	1·25
	1·00	0·41	0·39	9·50	0·82
	10·0	0·04	0·04	0·95	0·08
	20·0	0·02	0·02	0·48	0·04
At −20°C	0·2	1·94	1·86	45·20	3·92
	0·33	1·18	1·13	27·39	2·38
	0·66	0·59	0·56	13·70	1·19
	1·00	0·39	0·37	9·04	0·78
	10·0	0·04	0·04	0·90	0·08
	20·0	0·02	0·02	0·45	0·04
At −30°C	0·2	1·87	1·79	43·41	3·76
	0·33	1·13	1·08	26·31	2·28
	0·66	0·57	0·54	13·16	1·14
	1·00	0·37	0·36	8·68	0·75
	10·0	0·04	0·04	0·87	0·07
	20·0	0·02	0·02	0·43	0·04

Based on $\lambda = kT/\pi D^2 p$, where λ is the molecular free path (m), k is the Boltzmann constant, T is the absolute temperature (K), D is the molecular diameter (m), p is the pressure (Pa).

The concept of a molecular free path likewise is a device to help visualise the probability or otherwise of another molecule (*of the same kind*) being within effectual range. Whether treated as perfect gases or not, sufficient comparison can be visualised by comprising models (since sufficiently accurate and detailed measurement is by no means easy in practice) from Tables 11 and 12 to enable some general statements about the nature and style of factory-scale equipment to be made.

For instance, leakage air which has lost internal energy entering the chamber may gain appreciably in temperature passing heating shelves, only to be cooled down as it penetrates a system of refrigerated coils to get to the non-condensables extraction pump. The subliming vapour, on the other hand, at saturation temperature corresponding to the ice to start with, works its way out of the foodstuff past a warmer dried layer, dropping in pressure but gaining in superheat as it does so, past more warm metalwork, and mingles with air of ponderable temperature. Together they advance as far as the coils or whatever shape the refrigeration system adopts to enslave the condensable portion and *successfully* prevent it from proceeding further.

PRECIPITATIVE LANDING

By way of descriptive illustration (other uses will occur to the reader), in an expanded metal sandwich for instance, where the foodstuff ice temperature might be at −7°C, the heating plates well above 60°C, the vapour temperature on emergence from the sandwich early on in the cycle is then say at +3°C, having dropped fractionally from the subliming ice front saturation pressure of 0·66 mm Hg to around a (partial) pressure of say 0·5 mm in reaching the cabinet void proper. Its mean velocity could then be around 17 m/s and its molecular free path (m.f.p.) about 1·95 mm.

Unavoidable chamber leakage might maintain the partial pressure of air intermingling with this vapour in the void at around 0·2 mm, possibly warmed to 10°C, with a mean velocity of perhaps 0·3 m/s and m.f.p. of about 2·17 mm. They effect each other's temperature as they are drawn to the extraction area. The 'spot' chamber pressure is of course the sum of those two partial pressures.

While some 11% more vapour molecules than air molecules are then likely to be confronted by, and to be striking, the refrigerated 'ice condenser' surface per unit time, they will be doing so with around 35 times the momentum.

On the bare chilled surface around −30°C say (saturation pressure 0·18 mm Hg) the vapour will 'stick' as tiny crystals endeavouring to achieve dendritic open-work form depending on the intensity of arrival of further (mixed) molecules. Beyond this area, the air molecules are both cooler, i.e. have a smaller m.f.p., and largely on their own.

Later in the cycle, the dried layer initial part(s) of the vapour path becoming greater in depth and complexity, the vapour partial pressure on emergence will tend to have dropped, although its temperature may well have risen to nearer that of the heating plates. The quantity too will have fallen off, but the unit volume will have risen. Consequently the mean velocity will perhaps be about 1·5 m/s now in the chamber, and the m.f.p. will have risen to around 5 mm. The steady leakage condition will not have altered all that much.

At such a stage then, the ratio of vapour to air molecules striking the 'ice' surface will have fallen considerably, as will their relative momentum.

Note that the foregoing is only one of many models that may be conjectured for the sake of aiding visualisation. While the figures may vary, their relationships at various parts of the freeze-drying cycle will yield cohesive patterns from which to study the 'ice condenser'.

During the cycle, or between defrostings, the open-work ice-crystal fragile surface will have accepted more vapour landings, have grown, and its perimeter risen in temperature, exhibiting an m.f.p. not much shorter than that in the approaching vapour, but still longer than that of the non-condensables. There is now more expectancy of an air molecule striking a particular part of the condensing ice-front than of a vapour molecule.

The front of the *snow* build-up will over a period have tended to compact, enabling the temperature of the front more nearly to image the parent cooling fluid temperature. The air, cooled and slowed down in negotiating the chilled system vapour trap, will tend to carry the latest vapour molecules (and some of the dendritic earlier ones) further, especially if surged. The system has to be engineered with a mind to protecting the vacuum pump(s) directly from surging vapour.

Typically, single type molecule theoretical reasoning and common sense suggest that, if a gas has to traverse a round pipe at optimum flow, then it is better to have the tube relatively large in diameter and short in length. Conversely, if maximum ice settlement and formation is required, the obstruction should be severe and continuous, and the vapour flung or carried forcibly against it, although perhaps not with such vigour as to dislodge what is already there.

Standard forms of ice-condenser range from parallel flat vertical edge-on hollow plates, either alongside the sublimation area or in a supplementary chamber, to finned tubular batteries, the latter either lining singly the wall and end of a cylindrical condenser housing, or filling a rectangular cavity *longitudinally* such that entry is towards one end of the set of tubes and non-condensables suction remote at the other. Lesser use has been made of batteries of cross-sectional complexity, where the 'air' has to meander between offset rows of transverse finned coils, although this is a style of chiller favoured in cold-stores, for instance.

Obviously treatment of the two types of gases is at cross-purposes; a method has to be found which handles both effectively at all stages of the drying.

The rate of diffusion, G_m, of (water) vapour through *stagnant* air to the (ice) condenser for plain ice sublimation is quoted by Mellor (1978, p. 20) as of the form

$$G_m = \frac{MDP}{RTl} \ln \frac{P - p'_c}{P - p'}$$

provided that $p' \neq p_c$ when $P \neq p'$, where D is the diffusion coefficient of vapour in air, M is the molecular weight of vapour, P is the total pressure, p' is the partial vapour pressure outside the interface, p'_c is the partial vapour pressure outside the condenser, l is the distance between the 'evaporator' and 'condenser'.

He then takes average values of $\bar{D}$ and $\bar{T}$ since 'there is a temperature gradient'. Then he avers that, since DP is independent of P and proportional to $T^{3/2}$, the form reduces to

$$G_m = \frac{k_3}{l} \ln \frac{P - p'_c}{P - p'}$$

where $k_3 = M\bar{D}P/R\bar{T}$ varying as $T^{1/2}$ *but* can be considered constant.

From this one can understand why, since the sublimation rate varies apparently inversely as the distance between the evaporating surface and the condensing surface, many designs arranged these surfaces as closely adjacent as possible, even when the partial pressure of non-condensables near the first surface $(P - p')$ is not very different from the partial pressure of non-condensables near the second surface $(P - p_c')$.

As long as there is no non-condensables pump, the quotient $(P - p_c')/(P - p')$ would be greater than if there were, and the non-condensables would not remain 'stagnant'. In this hypothesis also the non-condensables are apparently unaffected by the difference in temperature, so one is bound to conclude that it applies merely to a very small model system. Certainly it would not pertain in (say) a tunnel with a car of subliming material at one end and a single operating ice-condenser at the other, physically many metres away, regardless of the state of absolute pressure maintained by non-condensables and drawn before sublimation was started. Such a system is difficult to isolate because of the lack of heat lagging on the enormous expanse of tunnel. It would be easier to simulate in a cold store at total pressure approximately atmospheric, especially since the ice-condenser/evaporator coils are usually near the roof and the 'thermals' aid and abet.

In the 'normal' freeze-drying which Mellor commendably describes in his 'fundamentals' book, the low total pressure in the chamber means low partial pressures of (other than water vapour) gases in the dried-out 'pores'. Heat transfer, hence the heat for sublimation, is assumed to be supplied mainly through dried fibres or through still-frozen layers (depending on the type of heat source, shelf or double-contact or finned tray). Thus conduction is principally involved, and heat input depends on the conductivity of the particular dried fibres (generally low) and the ice, with the precaution of not melting ice next to the heat path tray. Of these, ice has by far the higher conductivity.

In such a system there are relatively few gaseous molecules, other than water vapour molecules, in the dried-out labyrinthian spaces; the extraction means gives them little reason to remain there. The water vapour molecules which are next to the ice-front will be at the saturation pressure corresponding to that ice temperature, which can be measured fairly accurately. It can at least be distinguished from the main body of ice temperature.

Conditions between the ice-front and the exterior surface of the material being freeze-dried have been a matter of considerable hypothesis and conjecture. In many instances passage to the outside will be through warmer layers, so the vapour becomes superheated, increasing in volume at the same pressure or at a lower one. Either way it will emerge to encounter, or more strictly share the space with, relatively more (air) molecules because of leakage. Together they will contribute to the absolute pressure in that part of the chamber, but their progress from it may be relatively independent.

When the total pressure in the chamber is deliberately and rapidly raised by injecting atmospheric air or other gas(es), these will rush in with a velocity at the expense of their internal energy and the work they do on themselves. What happens next depends largely on which system of freeze-drying is in operation.

Almost certainly some of the in-rushing gases will encounter parts of the drying chamber such as the product heating system which will be relatively very much warmer. They will, if not creating actual shock waves, eddy considerably and, for the first part of the in-rush at least, pass over these hot surfaces at speeds that will allow heat transfer to them to be highly efficient.

This will increase their own internal energy (as they slow down) perhaps beyond what it originally was. Their activity when they penetrate the interstices of the partly dried foodstuff will be all the greater. There will be a slight delay while heat transfer there takes effect on the ice. Evacuation of the interstices (at a much slower rate than the filling of them) may have begun before evidence of any effect on the *total* pressure change rate is observed. Cycling is recommended (Mellor, 1978) to be quite rapid, with a dwell at the upper pressure of perhaps one-third the cycle time, which must therefore depend on the inlet aperture and the pumping capacity, among other factors which themselves are variable. Energy justification comparisons should not however be confined to refrigeration and vacuum pumping kWh, but must pay heed to the added utilisation of the heating system by the in-rushing gases.

The high initial speeds reached will also affect the 'ice condenser', the evaporator 'surface' ice, or rather the 'snow' load. Some of that already deposited will be blown off, to be borne (depending on the conditions) further along the surfaces or towards the mechanical pump(s) for non-condensables. This will make the surface and its temperature more receptive and tenacious to further sublimed vapour. The initial blast will not 'hiccup' a Rootes blower, or indeed a backing pump provided a suitable oil is chosen, which does not back-stream to any extent.

Observation and confirmation of these happenings require adroitly placed windows and temperature sensors in the chamber space. They are awkward to quantify, which is why much mathematical theorising based on stylised modelling has persistently been aired. The variation in the rate and timing of sublimation has to be inferred, and only overall assessments of time to a particular dryness, i.e. commercial criteria, have been experimentally compared with sufficient certainty.

The explanation of the benefit of cycling as described by Mellor would be independent of the means of extraction.

Long before 1963, when Leybold had installed in Italy a 2-m diameter sliding door between sections of a tunnel drying chamber, that company had hydraulically operated fast-moving large butterfly valves between the product tunnel and the ice-condenser(s). They were thus able to measure the rise in pressure in the chamber at various starting points not only when there was no product in the chamber but also when dried, partly dried and frozen or unfrozen 'new' product was being tested, say for vapour flashing. It must be remembered too that, because of the type of finned dish that was peculiar to their design, any influx of atmospheric air (or for that matter nitrogen) had to find its tortuous way around most of the heating and heated surfaces before reaching the product.

During tests carried out on a Leybold food-processing freeze-drier in Holland in 1960 by MAFF, it had been observed that the ice on the evaporator coils changed pattern after a rapid release of vacuum. Obviously the scantlings or relative dimensions of even a series of freeze-drying units will affect the actual behaviour, but it may be generally stated that, from such contrived rapid influxes, the proportion reaching the product will have been warmed, and that part initially (and perhaps inadvertently) directed towards the ice on the evaporator will affect the disposition of that ice, or at least its snow-like precursor. It is perhaps a reflection on the state of the art of design of such systems that consideration of these effects has not been over-manifest!

While the specific heats (at constant pressure/temperature) of air and pure nitrogen are very close, and their adiabatic indices are therefore also close (Table 13), such average values are appreciably different from those for helium, an academic and expensive gas. With water vapour, variation of the specific heats is affected by pressure as well as temperature, and an *average* value has little meaning. At the conditions outlined it is perhaps more like CO_2 in behaviour than the others.

For the purpose of visualising practical numbers, assume a chamber

TABLE 13
Mean Values at 15·5°C and 760 mm Hg

Gas	*Specific heat at constant volume*	*Specific heat at constant pressure*	*Difference*	*Index*
Air	0·172	0·241	0·069	1·40
Nitrogen	0·178	0·249	0·0707	1·40
Carbon dioxide	0·157	0·202	0·0448	*c.* 1·29
Helium	0·768	1·265	0·497	1·65

of 10 m^3, the particular *total* pressure within it averaging 0·2 mm Hg (virtually all water vapour), raised to 20 mm Hg periodically by the deliberate ingress of (dry) atmospheric air at 20°C through a 5-cm diameter instantaneous orifice. Assume also that the ice temperature is −20°C (dry saturated vapour pressure 0·77 mm), there is appreciable product already dry, and close to the heating surface temperature of (say) 40°C the surface dry layer may be restrained so far on in the run. The fall in pressure from 0·77 to 0·2 mm plus the inevitable superheating will have increased vapour volume by $4\frac{1}{2}$ times that at the ice-front. Mass flow will be the sublimation rate; the velocity of emergence will depend on the ratio of external interstice area to ice-front area.

The refrigerant at −40°C (just under 0·1 mm equivalent) may be keeping the 'snow' on the coils at −35°C (0·168 mm) when the cycle starts. The velocity of the air jet entering the chamber is around 700 m/s, falling to 600 m/s as the first rush of air is compressed again 100-fold to 20 mm Hg. In some 15 s about 1/3 kg of air has entered the chamber and swirled around it. Some of this air will shoot over the evaporator coils right away, will be colder than them, and will entrain snowy deposits en route for the blower. Some will eddy around the heating platens and rapidly regain temperature, since the heat transfer coefficient will be favourable. At entry valve cut-off the air will still be in random motion and, although the valve may have been actuated by a chamber total pressure signal at 20 mm, the air will not have completed packing the interstices to that value.

If a partially dried product is used to determine how long it takes to exhaust the air in it from 20 mm Hg to 0·2 mm Hg this can form a guide as to how long it is likely to take to fill the same interstice spaces over the same range—quite a large proportion of the recommended dwell time at the 'higher' level. The popular assumption is that a combination of two features occurs. The air brings heat to the ice-front via itself and by warming the dried part of the product, and the vapour in the dried part is compressed against the ice-front. In adjusting to its new state it sublimes the ice more rapidly, building its partial pressure to produce a vapour flash. By this time the partial air pressure is on the wane and is temporarily overtaken. The vapour on emergence finds the coils more inviting. The advantage will continue while the dry portion retains its improved heat supply.

The system or suggestion of employing air (non-condensing) molecules as a means of supplying the heat of sublimation through the 'pores' as well as the 'fibres' of the material being dried will depend for efficiency on the heat receptivity of those 'fibres'. The ice 'front' is, as it were, embedded in these fibres and will respond to energy transfer in some sort of in-depth way via them as well as a direct frontal attack. The ice molecules being excited to leave their solid state in a matrix of soluble solids of pure coffee (say) would not have quite such an advantage.

One line of experimental investigation which lacks reportage, with existing refrigerated-condenser and backing-pump systems, on a commercial-scale level is when, after loading and drawing a vacuum to remove the air and with it the oxygen, replacement proceeds with nitrogen to atmospheric pressure. Drying then takes place on the partial pressure of water vapour only, in isolation. One might infer that oxidation of dry layers would be avoided, ample heat would be supplied through the nitrogen, and at the end of drying, which might have to be slowed down (?), the pores would be full of inert gas. The heat supply rate has to keep the ice sublimation temperature at a steady level; while this is relatively easy to control in a frozen-on-shelf situation, one has to employ a different strategy in frontal attack contraflow.

Refinement of the method has not received as much attention as perhaps it should; ingenuity's reward would be the avoidance of having to use additives to retard oxidation, with carrots for example.

The foregoing critique puts to the question a number of well worn treatises. Nevertheless it is not by any means dismissed as a well-nigh impossible academic exercise to construct a *universal* model, mathematically based, as a forecasting tool for process optima, without introducing factors that themselves are highly variable, indeterminate and particular at most to a 'fuzzy theory' structure. Such calculations are most suppositional in treatment of cell contents, marvellous enough in functional activity, and still sources of value in digestion and ultimate assimilation.

Chapter 7

Factory Features Peculiar to the Process

FACTORY STYLES

Pharmaceutical freeze-drying installations working with products dried in their phials, which are stoppered prior to emergence from the cabinet(s), often have the plant built into a wall, so that the front may be clear of 'engineering accretions'. This implies a 'clean' room for loading and unloading, and it was an understandable extension of previous practice, when other materials became practical possibilities, to treat food plants in much the same way. Indeed, where pre-freezing is the exclusive routine, and rapid filling and emptying is aided by produce transport on rail or car, then single full-door cabinets sequentially operated will decide the format and centre-piece style of the factory (Fig. 26). In Oregon, for example, 14 such cabinets are in line.

More likely, since the product has perforce to be dealt with more rapidly than the produce, the emptying and packaging sequences tend to demand a 'through' arrangement, and therefore may call for a second full-size door to the rear or exit end, from which it follows that many of the accoutrements will be disposed to one side. At least one (Italian) plant mounts its substantial ice-condenser on the roof of the cabinet, whereas a Danish design, more frequently defrosted, stows a pair underneath.

Steam-driven vacuum augmenter systems (see Fig. 33) have stages with the suction port generally at right angles to their main axis in each case; thus they tend to take up a lot of headroom. Granted no barrier to height, they can be arranged in common support with a barometric leg on the elevated vapour condenser, but ordinarily one might expect

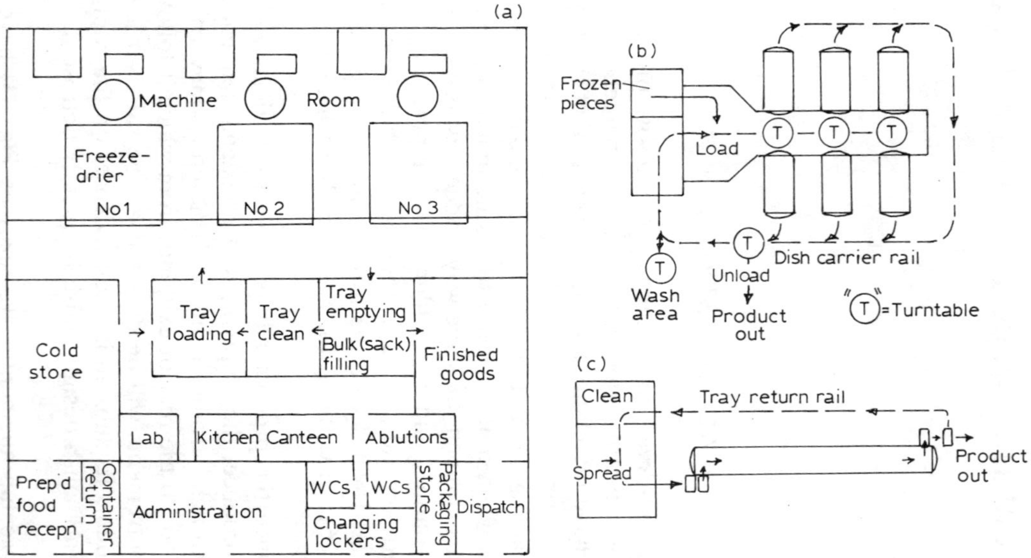

FIG. 26. Elements for a contract freeze-drying factory (material brought to it semi-prepared). (a) One-door cabinets loaded by hand with large trays. (b) Two-door cabinets loaded by carriers of dishes. (c) Tunnel drier with individual tray transport.

them to be housed mezzanine fashion. Applied to individual cabinets they probably dictate spacing, but commendably neat positioning has been achieved with a multi-stage communal set embracing the vapour discharges of four large batch cabinets (Irish Sugar Company, Mallow) harnessed in such a manner that their blended outputs formed more of a rippling load, if not an entirely steady one. None of the variations in layout of this type of vapour removal gear debars the cabinet(s) from having doors at either end.

Whilst continuous (tunnel) plants have been envisaged as being of horseshoe or hairpin shape, with a bend in the middle to bring the discharge station back to the same plane as the loading station, actual ones so far have all been straight, so that, depending on the length of the process in time measurement, the exit is physically positioned some distance from the entry point. Designing the cross-section to be larger makes no difference in this respect.

It can be seen from the aforesaid that the design of the actual freeze-drier(s) is a major influence on the layout around it/them. Only very large organisations depart on expansion from keeping to the

FIG. 27. Right- and left-handed drier pairs in line abreast before being built-in (New Zealand).

original supplier. Duration of the process is also significant; if attendance on the machine, lasting perhaps 1½ to 2 hours, is once per 24 hours, then a single shift of servicers can skilfully deal with 4 or 5 cabinets in turn (Figs 27 and 28). An extended shift of the same few can possibly deal with the preparation and disposal sequences also. Some contract freeze-drying companies receive the produce in already frozen but free-flowing form, and likewise simply bulk-pack in multi-wall sacks for onward despatch to a consumer-portioning ultimate blender, packer and distributer, who may or may not be the originator. An arrangement of this nature suits the dehydrating company very well, because its specialised capital investment can be

FIG. 28. Interior of batch drier showing heating shelves on tandem racks with hose connections. An external car is brought up to exchange them rapidly.

continuously and fully utilised regardless of season or availability of (or demand for) any particular commodity, while the various users obtain a more even cost service to their particular specification without the proviso of protracted capital risk. It cannot be said that any dearth of entrepreneurs has curtailed expansion along these lines, but what has tended to result is that a considerable market in freeze-dried ingredients of soups, yoghurts etc. has conveniently gone 'underground'.

Harvesting, sorting, sectioning and general preparation (Fig. 29) for freezing (and/or drying) of fruits and vegetables are as normal for a *freezing-only* operation, *but* the actual quick-freezing, while using standard equipment, will be at more appropriate speeds and levels of temperature. The raising to dehydrating temperature may be carried out for the most part in a 'holding' room and with the material spread on the drying trays or in dishes and rack cars ready for shipment into the dryer.

The same distinction can be said for the butchering, dissection and preparation of all kinds of meat pieces, but these treatments will be carried out (as in other meat plants) in rooms clearly designated for these purposes and scrupulously segregated in the normal manner. Thus it is feasible, and indeed likely, that a general purpose installation will have several modes, routes and systems leading to the frozen state preparatory to the logistical feeding of the drying hall and driers.

Liquid concoctions, soups and beverages of all kinds are specialised in their preparation and concentration stages, although they may all be brought to the solid state and particulated in much the same sorts of machines.

Batch driers, as the name implies, deal with one fixed quantity at a time, and that usually completely. The produce is arranged to be static while the (whole) process is performed, environment variables being altered around it to suit.

The 'continuous' description applied to large freeze-driers refers to sequence batching in the main, where trays or dishes rest throughout stages of the process, each stage being in an assigned part of the drier where the environment is set principally by the heating temperature. Physical entry and discharge are effected by means of locks, capable of dealing with car-loads or single trays (Figs 30 and 31).

There have been several designs built of truly continuous freeze-driers, some of which are still producing strongly, all centred on

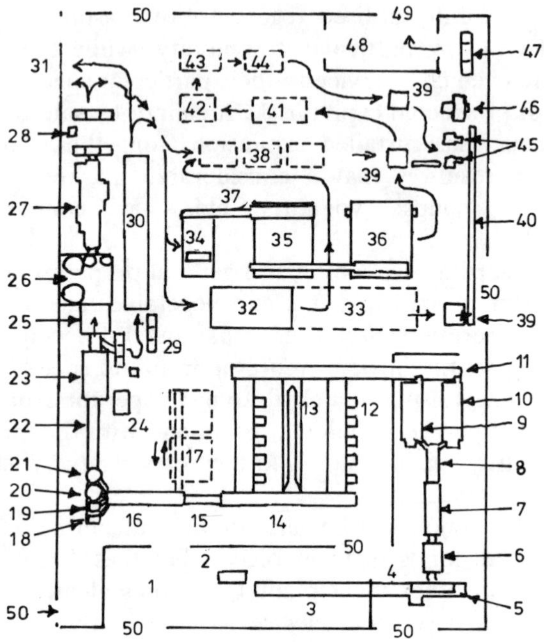

FIG. 29. West Bay, Experimental Factory, showing typical unit layout for fruit and vegetable preparation for drying, and incidentally the unit layout for hot-air drying and packaging. 1, Raw vegetable store; 2, grader; 3, flume; 4, stone trap; 5, wheel elevator; 6, pre-washer; 7, inspection belt; 8, elevator; 9, steam *or* flame peeling; 10, abrasive *or* caustic peeling; 11, conveyor; 12, trimming belt set up for leaf vegetables; 13, trimming belt set up for root vegetables; 14, conveyor; 15, elevator; 16, conveyor; 17, buffer storage; 18, shredder (USA); 19, shredder (UK); 20, stripper/dicer (USA); 21, stripper/dicer (UK); 22, elevator; 23, stripwasher; 24, starch recovery; 25, water blancher; 26, dosing tanks; 27, cooler; 28, weighing station; 29, alternative weighing station; 30, steam scalder; 31, runway to 'vacuum' drying; 32, hot air belt drier; 33, extension hot air belt drier; 34, hot air cabinet drier No. 1; 35, hot air cabinet drier No. 2; 36, hot air cabinet drier No. 3; 37, warm air ducts; 38, hot air bin driers; 39, container filling; 40, gravity conveyor; 41, preparation for pressing; 42, presses; 43, guillotines; 44, treatment after pressing; 45, container sealing; 46, gas packing; 47, container testing; 48, lacquering and sample identification; 49, transport to constant temperature stores; 50, access doors.

produce in small particle form to make operation of the vacuum locks easier. The earliest suggestion was one by Greaves (1962) which involved a frozen plug being advanced slowly into the chamber, while a rotary scraper inside tore or shaved it into a particulate form in which it would rapidly dry. Schemes to flash-freeze by evaporative chilling of liquid injections to the vacuum space led to complications of blockage-freeing devices with built-in uncertainty.

FIG. 30. Dish carriers on return rail alongside tunnel. Vacuum lock takes 126 dishes at a time.

Later systems pre-granulated the material, transporting it from a sealed hopper into the vacuum space (sometimes in dollops) by a rotary valve. Therein it had to be spread out evenly, to expose as much surface area as possible, while presenting opportunity for penetration of the heat required for sublimation. Relatively deep beds were formed on a succession of swept shelves or slowly moving belts or stainless steel bands.

Fluidised beds were more popular, where vibration is used to shuffle and gently throw the material onwards. Both vertical and horizontal forms (the latter especially) have been used for coffee. Because of the short drying times, particle size distribution has to be very close, 1–2 or 1·5–2·5 mm (say); recycling of fines occurs far more than in static drying.

FIG. 31. Twin locks take one tray each for/from the internal racking system. These trays are elevated to between the heaters in vertical array, and used to push each line of trays along.

Freeze-dried raw beef in powder form was displayed at ANUGA in 1987, for use in sausages, packaged soups and infant food; powdered banana and vegetable from the USA were also shown.

Cryo-Maid of San Leandro, California, pioneered in the early 1970s a continuous freeze-drier (Fig. 32(b)) claimed to process ice-crystal-controlled semi-liquid and discrete particle foods on a shaker system at more than four times the rating of equivalent-shelf-area cabinet freeze-driers.

Preparation involves concentration, chilling, then spreading on a stainless steel belt, which itself is chilled by spraying trichloroethylene on the underside, in three zones set at −32°C to −68°C. The frozen layer is fractured, broken, reduced to a classified particle size before entering a (jacketted to −38°C) hopper, gate controlled to feed the material regularly into the vacuum chamber, which is held at a nominal total pressure of 0·05 mm Hg by a well baffled pumping system.

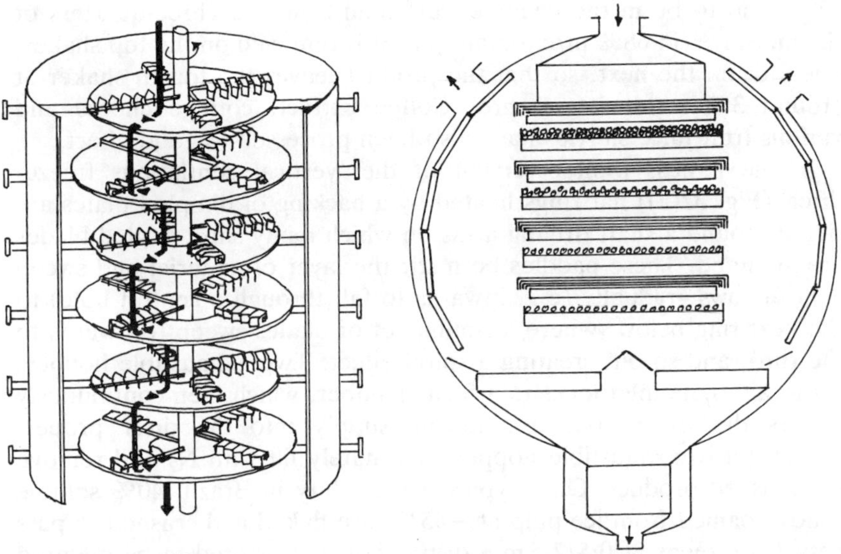

FIG. 32. Continuous freeze-driers. (a) Krauss–Maffei (swept type). (b) Cryo–Maid (vibrating type).

Four conveyor decks are dynamically balanced on an internal frame, each being stepped so that the particles tumble along as it shakes on leaf springs. Each deck is cooled to −38°C by an internal refrigerated labyrinth. Interspaced with the stack of vibrating 'shelves' are radiant heaters, shielding what is above by a stainless steel sandwich of multi-layer heat baffles, so that their energy in six zones is directed downwards to the dancing particles proceeding in reverse direction directly below. Heating fluid temperature ranges from 315°C at the start to 200°C focused on the bottom layer. Product is eventually directed towards an exit hopper serviced by locks.

The sublimed vapour is condensed on six pairs of five refrigerated (−68°C) panels, built along the walls of the cylindrical freeze-drying chamber, in such fashion that periodically, when heating fluid temporarily is caused to invade the panels, the sheet ice so far formed slides off to land on a pair of longitudinal conveyors. These transport the sloughed ice to an internal crusher, lock and separate exit hopper, where it is melted before discharge to waste.

All sequences and settings are automatically controlled, making for a highly sophisticated set-up. Material is claimed not to rise above

27°C, and to be in the chamber less than 2 hours. Three-quarters of the moisture (in 68% m/c tomato paste) is removed on the top shaker, one-fifth on the next, so that the product leaves the fourth shaker at around 3% residual moisture. Coffee extract, cottage cheese and various fruit juice slurries have also been processed to good effect.

In the Krauss–Maffei version of the 'vertical continuous freeze-drier' (Fig. 32(a)) flat rings heated by a backing of dimpled plates are tiered around a shaft driving arms on which many short-angled blades are mounted. These paddles both stir the layer of material and sweep it in furrows gradually (e.g. inwards to fall through a central hole) to the next ring below, where a similar set of blades sweep (outwards to the rim), and so on, creating a spiral effect. Two exhaustible hoppers act as alternate inlet locks to a feed chamber, which then continuously meters the frozen but free-flowing supply into the drier proper. Similarly two exhaustible hoppers alternately fill with N_2 and remove the finished product. On a typical coffee duty in Brazil, 40% soluble solids, foamed from ice-pulp at −45°C, are flaked and crushed to pass between screens at 0·5/2·5 mm distinction, then stored to be metered into the drier.

The chamber operates in the region of 0·5 mm Hg. The first 2 plate rings are at 40°C, the next 6 at 67°C, then 21 at 74°C and the last 6 at 35°C. The twin ice-condensers operate between −40°C and −50°C, switching to defrost. The mean time of spiral transit over the heated surface (175 m^2) is around 100 minutes, leading to an overall sublimation rate of more than 500 kg of ice per hour. Vapour volume is such that entrainment of fines into the vapour discharge ports is appreciable. Very large cyclones or baffle chambers are therefore required to act as dust separators between the drier proper and the ice condenser. Recovered fines are collected from the conical base of the separator and conveyed by screw elevator to the start of the last fifth of the plates, i.e. below the port(s) leading to the separator, and into a region where less vapour is being emitted. Recirculation ratio is unknown. Resistance thermometers, trailing from the paddle arms, serve to provide some indication of product dried surface temperature! Feedback of information from all over the plant has enabled a high degree of automation to be instituted. The heyday of such plants was 1968–73.

Simpler deflector plate arrangements *within* the drying chamber, particularly for arresting by collision fines from vibrating systems, have been tried or mooted. Provided the particles do not shatter further (and have the awareness to accept in which direction the arrow-head is pointing), all such schemes reach a success limit. Teflon coating of the baffles will not necessarily increase this.

It is in the 'after-process' department of 'dryness preservation' that most differences from other types of factory may be observed. Largely depending on the method of releasing the vacuum at the end of drying, the rate there at which a protective environment barrier has to be sealed around the dried material will have to be established. If commercial nitrogen is used to break the cabinet atmosphere, the accompanying twentieth by volume of oxygen will also have invaded the tissues, perhaps affecting the lipids in the product but virtually making no change in the ERH. The atmosphere of the unloading room at least, however, has to be tolerably comfortable for operatives, as well as presenting as little as possible hazard to the product (unless emptying is conducted by remote control in a N_2-filled room, as has been envisaged, and indeed built, but found expensive to use).

Work-force personnel themselves produce heat and moisture in the air, and do not take kindly to draughts. Therefore a negotiable balance has to be struck, with the air conditioning set to give perhaps as low as less than half an air-change per hour, and maintaining a relative humidity at a slight over-pressure, of (it has been known) 35% at 20°C. This is equivalent to 0·005 g moisture per g of bone-dry air, or 1 g in 170 m^3.

For example, at this temperature and level of humidity in the air, freeze-dried raw fish (*if allowed to*) would equilibrate at a moisture content (dry basis) of around 10%, cooked pork or chicken around 9%, fruit or vegetable pieces around 6–7%, tea around 8% and coffee say 3%. But it would be folly to allow them to do just that, and in fact the situation demands hasty apportioning and packing off in a very few minutes to ensure less than 2% moisture content (dry basis).

Broadly speaking, their drying end-point and release temperatures will be above the room temperature; such product temperatures, if not allowed to fall significantly by delay, will be an important aid to prevention of a more rapid take-up of moisture.

Not surprisingly, freeze-drying factories have not yet achieved a standardisation in physical style, such as that in a heat-preservation-by-canning factory or a straightforward freezing factory. Nevertheless, custom-built as most are, even by make of drier, they exhibit a certain panache which makes viewing them an engineering pleasure as well as a food technologist's inspiration. This may in part be due to their evolution as individual ventures without the umbrella or shared confidence of a particularly devoted trade association.

CONTRACT PACKERS

While freeze-drying factories under contract prefer to deal in bulk, disposing of the dried produce in (say) multi-plastic-liner sacks, there exist convenient intermediaries who will carefully transfer the materials to individual portion or handy catering size packs. From a range of sources, ingredients may be blended into special meals, being formulated according to demand.

Given that their basic ingredients are delivered in bulk, their special skills tend to be associated with compilation, combination and final retail packaging in a controlled working atmosphere which is designed to prevent undue moisture pick-up, but still permits of comfortable working conditions, all towards an acceptable finished goods storage life.

The pulmonary process demand for oxygen, and exudation of moisture, means that the packing room personnel need air regularly changed, with freedom from draughts and while subject to strict humidity control. A compromise condition between what is amenable for the packers and sufficiently 'non-adverse' in the time scale of exposure for the product is 20 ± 1°C and 30–40% RH, with 0·007 g/g as the maximum absolute humidity.

The room air changes per hour will depend on its size, lighting, the number of people, their activity (individual heat input can vary from 175 W typically to 225 W) and the type(s) of machinery enhoused; air movement of around 1 m/s should be tolerable.

Part III

EQUIPMENT

Chapter 8

Functionary Parts

MASS TRANSFER/VELOCITY SYSTEMS (Fig. 33)

Using the kinetic energy from steam to entrain the space contents, evacuation of the freeze-drying chamber in factory-scale installations can be by booster two-stage ejector direct to atmosphere via a silencer. Modern ejectors can provide a pressure difference of 15:1, so two stages on steam at 175°C or above can easily bring the total pressure of a fairly large chamber down to below 4 mm in under 3 min, and below 1 mm in under 5 min, assuming that the foodstuff load is pre-frozen.

Commercially tight systems should have an air leakage value† of not more than

$$y = a \,.\, x^b \tag{1}$$

where y is the air leakage (g/min), x is the volume of chamber (m^3), $b = 2/3$, $a = 16$ at start of pull-down (i.e. atm. pressure), 4 at 3 mm, 2 at 1 mm.

Thus a chamber of 10 m^3 should be constructed tight enough, when there is no pressure difference across its seal, to draw only 75 g/min of air, and when the pressure difference is 760–3 mm, say, to draw only a quarter of that.

The greatest concentration of the leaking mass of air is independent of the condition inside the chamber and is given by

$$\frac{W}{A} = \frac{v}{V}, \quad \text{where} \quad v = \sqrt{\frac{2gm}{m-1} P_1 V_{s1}} = 311 \text{ m/s}$$

† A leakage of 1 litre/s at a pressure of 1 micron inside and 1 atm outside is termed 1 lusec and is the (high) vacuum engineering unit referred to the pumping means. Torr-litres/s is also used. All units are mass rates.

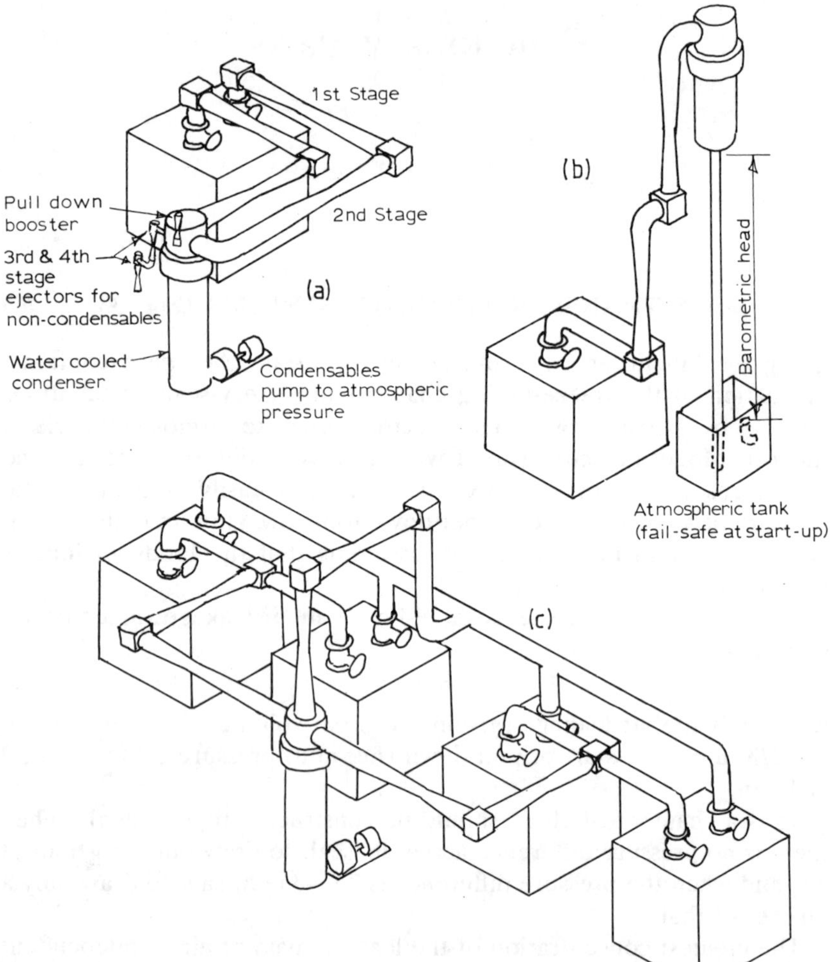

FIG. 33. Arrangements of steam ejectors for augmenting vacuum. (a) Sets for (2/3)/(1/3) maximum load; low condenser backed by pump. (b) Barometric leg dispenses with need for pump. (c) Manifold system for several cabinets on staggered cycles.

a formidable velocity providing, if the leak were localised, energy that could result in tangible force.

When the vessel pressure is around 1 mm for instance, then the velocity of the in-rushing air can attain more than twice the speed of sound as it penetrates the chamber, setting in motion shock waves which can visibly affect the materials inside. Should these be feathery forms of ice or snow, they may well be dislodged.

Whether or not the substance to be dried is likely to give off non-condensable gases, it is still usual to double the value obtained in eqn (1) when calculating the evacuation load. This air is also extremely cold—far colder (and in the first instance far drier) than anything else that pertains in the chamber. One must not underrate its properties.

Exhausting the chamber to 1 mm Hg in under 5 min, by projecting the air in it to atmosphere, demands an appreciable quantity of steam, perhaps 12 kg/m^3 of space. However, vapour need not be compressed to atmospheric pressure but only to as high an absolute pressure as will allow condensing water to exist in the system, absorbing latent heat, thus permitting a combined liquid discharge to be removed by other means.

The augmenters are convergent/divergent so that the combined flow of vapour slows down as it reaches the stage exit. The pressure and temperature that it arrives with is dependent on its nature, the amount entrained, the tunable steam supply and the physical design shape of the stage, of which the latter is fixed. There is a collapse of volume on meeting the 'cooling water' which of course has a saturation pressure corresponding to its temperature of the moment. Harmony of condensation will depend on this temperature, as the 'steam' temperature can vary above as well as mainly below it. This is why a direct contact jet-type condenser is preferable to other styles. Supply in abundance of natural cooling waters which can be discarded to waste are available cheaply in estuaries, and it has been traditional to site some freeze-drying factories near such sources for that purpose. These are, however, subject to seasonal if not more frequent variation, and as the temperature fluctuates so does the corresponding pressure. If, for instance, the water temperature rises, the steam supply to the augmenter nozzles must be 'kept up' so that the entrained vapour reaches the correspondingly higher pressure. If the water temperature is lower, tuning is not usually resorted to, and steam is wasted. Water is drawn in by the vacuum, and either a barometric leg or a liquid ring pump is used to bring it up to atmospheric pressure and out again.

Where mains water charges have to be met, the discharge water may be recirculated through an induced draught cooling tower. Depending on the wet-bulb temperature of the air (in the hottest months say 16°C in Europe and 24°C in the USA) for a loss of perhaps $1\frac{1}{2}\%$, each time round, the water can be fed to the condenser at between 21 and 29°C compared to say 13°C for estuary water. This means that the water has a saturation pressure anywhere between 11 and 31 mm Hg. The sublimed vapour load will vary or rather tail off quite noticeably during a run. It seems a reasonable benefit to avoid having a variable of this nature at *both* ends of the stage, and the provision of a particular, steady temperature of cooling water would overcome this.

At the *start* of an AFD run, for instance, the ratios (using published Aberdeen figures) of 'vapour of sublimation' to 'driving steam' to 'cooling water' are as $1:3\frac{1}{2}:330$, and *towards the end* as 1:150:46 000, simply because the feed of cooling water is dictated by the condenser temperature and the float, not by the 'vapour'.

Cheap and efficient *chilling* systems for water exist nowadays which can be set only to operate when the feed or return-water supply hints at rising a fraction above design temperature, while the flow can be regulated on the outlet temperature. The savings are to be weighed not against the steam saving (of which more later) but against the cost of inoperation! A *rigid* system depends otherwise on tuning the number of augmenters in operation by adjusting the steam supply to (or the number of jet nozzles working) a much more delicate and intricate procedure.

The entrainment load at any one time consists of, besides the water vapour, a (steady) leakage of air, and perhaps some trapped gases from the product base. As the run progresses, the proportions will alter, and so will the kinetic effect (Fig. 34). The adiabatic index C_p/C_v for superheated vapour at around 1 mm Hg is in the region of 1·29 while that for the driving steam may be about 1·23 and that for the leakage air could be taken as 1·4 say. Even reckoning on common behaviour by the throat, this measure would be perhaps 1·25 early on, falling to 1·23 at the end as the driving steam has less to entrain. Calculation of the throat area has thus to be a compromise, and the nozzles have to be adjusted on-site accordingly. The adiabatic index for helium, by way of comparison, is 1·66 and, although the practical value for calculation may be slightly lower, the standard formulae apply.

While the entry velocity of the steam to the jet or group of jets will

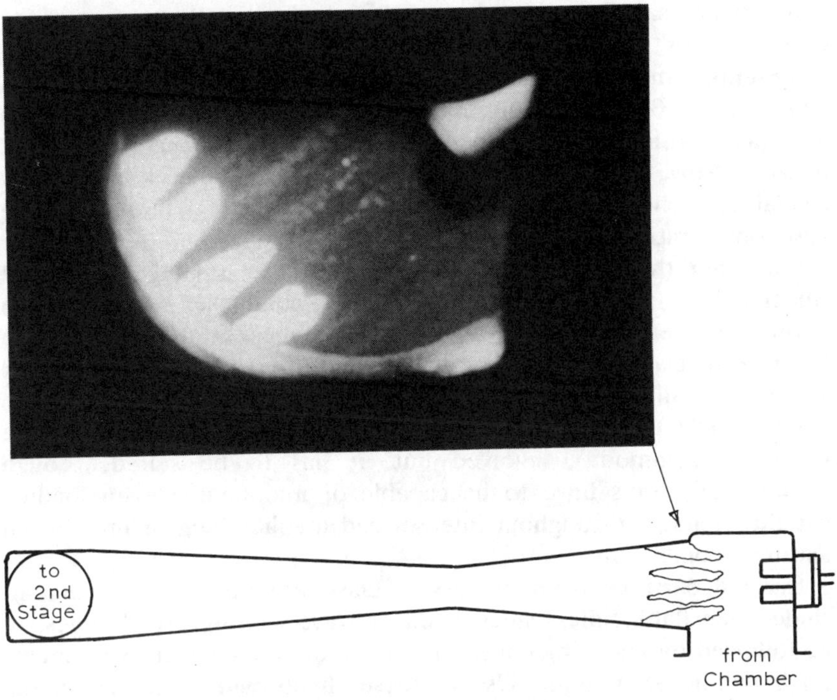

FIG. 34. Venturi systems are of finite capacity. Particularly in evaporative freezing, conditions in the combining tube of the primary augmenter may become such that 'horns' of condensed ice may develop. Superheated steam to the nozzles, or steam tracing near the throat, may be resorted to.

be negligible, the vapour velocity at the chamber ports (because of the enormous specific volumes at operating pressures) may be as high as 95 m/s, tailing to around 25 m/s. Large augmenter bodies are often of cast iron, as-cast and relatively rough inside. Machining is costly but gives an overall improvement in steam consumption of up to 5% (Hargreaves, H., 1985, pers. comm.). It is equally possible to coat the internal surface with glass, or at least pseudo-glaze it with a resin in the convergent and entry section. Understandably the so-called static layer next the wall of the entraining venturi represents more a proportion of the *load* towards the end of the drying cycle.

Several arrangements of steam ejectors in tandem/parallel have been tried or proposed, to be operated in a sequence which would allow steam usage more closely to match the load for most parts of the cycle, in steps.

Systems of augmenters stepped in ratios of 1/2 and 1/2, 1/3 and 2/3, and 1/7, 2/7, 3/7 have been discussed (Barrett *et al.*, 1964), although not much consideration has been given to cluster variation of steam jet nozzles. Now that computerised process control devices are readily available, such additional complexities should attract more running-cost consideration.

For other than barometric leg condensers, as a safeguard against steam failure, when suck-back might bring condenser water into the drying chamber, non-return valves are sometimes fitted. These have to be of a special lightweight type, not to be a strain on the ejector system, and allied to a vacuum air-break, or condenser isolator.

Receptacle design for foodstuffs depends to some extent on the particular commodity involved but, it has to be stated, though obvious, all forms have to be capable of maintaining even loading without spillage, throughout intense load manhandling or impersonal automatic operation.

Early departures from processing glassware on bare shelving included thin flat stainless steel sheets as trays with upturned edges, to be followed by matching coverplates as in double-contact experiments using ductile aluminium. Use of these sheets with expanded metal inserts (Fig. 35), first in springy stainless steel but later in more amenable aluminium, taught that the metal sandwich layers at 2·44 m by 1·22 m had to be pushed into the drier speedily *en bloc* from a tiered car which in turn had to be precariously loaded in a cold room. Brought down in later models to the handleable size of 2 m by 0·75 m, later 1 m by 0·75 m, these trays no longer flexed awkwardly. Moreover they could be stashed comfortably anywhere in a stack which stood as low as 0·5 m and as high as 2 m. The turned-up lip did no more than align the insert.

Loose material like cabbage shred, which is easier to spread before scalding, was found to be best contained evenly in a tray with definite strength in the rim, which was made from tube of round or preferably square cross-section, to which mesh of appropriate weave, perforation or tooling could be welded or otherwise affixed, free from crevices difficult to clean. These maintain their rigid shape to close tolerances and therefore can have automatic racking gear arranged to manoeuvre them.

A family of such trays emerged, for general use, based on the

FIG. 35. Expansion of slits in metal sheet produces a three-dimensional insert capable of aiding both thermal and vapour flow.

premise that the volume of material was best arranged for freeze-drying in a thin layer, up to 25 mm thick say, with a large surface contact area making up the cross-section, whether that shape comprised an arrangement of slab pieces or a bed of randomly loosely packed material. Some genuine endeavours were contrived to effect an artificial reduction in cross-section of Chateaubriand-style meat steaks, by impaling them on perforated hollow spikes projecting from the tray in relatively close order. These carried heat to, and vapour from, otherwise remote and therefore tardy areas, but scale-up proved difficult to implement.

A consensus gradually developed amongst equipment manufacturers, prior to any form of unanimity amongst the potential users, that economics of production precluded certain forms, thus directing development into general-purpose if somewhat blinkered channels.

TABLE 14
Comparison of Performance of Finned Dish Types for Tunnel Freeze-Driers (see also Fig. 37)

	Type 'a' suspended between heated shelves at 80 mm pitch	*Type 'c' suspended at 110 mm pitch*	*Type 'c' integral with shelves at 80 mm pitch*	*Type 'e' resting on heated shelves at 140 mm pitch*	*Type 'f' resting on heated shelves at 140 mm pitch*
Heating surface per unit chamber volume (m^2/m^3) (A)	12·5	9·15	12·5	6·85	6·85
Loading (kg/m^2)	8·0	15	15	28	26
Cycle time (h)	8·5	8	8	14	7
Initial drying rate ($kg/m^2.d$)	22·5	45	45	48	90
Overall rate to dryness ($kg/m^3.h$) (B)	4·7	6·8	9·3	5·5	10·3
Performance (B/A)	0·37	0·75	0·75	0·8	1·5

An alternative family of receptacles (Figs 36 and 37) arose from promotion of high-wall trays, originally manhandled at 0·75 m square, of 3 mm aluminium, with rims standing 7·5 cm high, for loose beds of vegetable material. For additional strength as much as an aid to heat penetration, internal spacers or rather ribs, quite 15 cm apart and of full height, were provided, welded in. Concurrently, wire mesh and/or expanded metal inserts, in the shape of internal dividers or partitioning 'walls', between the fabricated ribs showed promise of speeding up the process in the context of such depths of bed (Table 14).

Much closer partitioning of these types of dishes, which were subsequently reduced in overall base area, was achieved by extruded section, having end-plates welded on, with support lugs. Again this was now a design which lent itself not only to mass-production but to mechanical handling on trolleyed rack-cars. Also the option was open, as with the standardised trays, for external radiant heating, alone or allied with some form of contacting shelf.

The concept of a series of sandwiches, based on tiers of *horizontal* shelving supporting (by means of trays) relatively thin (up to 25 mm) layers of foodstuff, was thus seen to be contrasted by a series of sandwiches based on *vertical* partitioning (up to 25 mm) of a much deeper layer or bed of foodstuff—slabs being on-end, as it were.

FIG. 36. Type 'f' finned dish with inserts; see also Fig. 37(f).

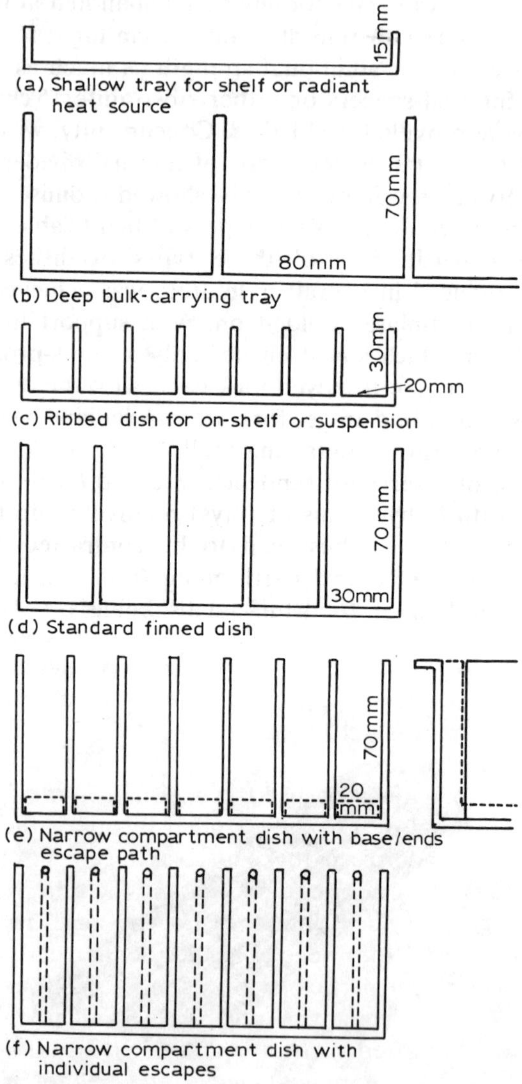

FIG. 37. The finned tray or dish; cross-sections of various styles of extrusion, and perforated inserts.

Although the cumulative heating shelf area was no longer a clear indicator of capacity, in the utilisation of the cross-section of the actual (circular) drying chamber, not a great deal of difference had transpired.

Some plants (notably in Japan) still use mesh fold-over, slip-on, sleeves to the ribs as an aid to vapour egress from particulate products, but most installers choose to concede the cause-and-effect argument of limiting demand for higher initial heating and vapour disposal rates. Had it been possible for the two major forms or design paths to journey more closely, the present predominant forms of continuous operation might have resembled a more parallel evolutionary style.

The Erin Foods freeze-drying plant at Midleton, near Cork, supplied by Stokes, is an example of an overhead-rail dish-carrier system where the ribs are not close, the dish is not so deep, and the carrier is remaining poised between the radiant heating plates as do the pushed trays in the Atlas system, rather than capable of being lifted, transported and set down for conduction contact as in the Leybold system. It is not a true hybrid but it deals, in sequential batch cabinet fashion, with a wide range of produce of the country at economical rates.

Lipton reported (Seltzer, 1960) looking at shelf-heated aluminium rigid trays, 45 mm deep, with partitions. At 70 mm apart, no difference from performance with open trays (using large meat chunks) was noted. At 45 mm apart, the drying time was reduced from 15 to 13 hours. The benefit was not linear.

Leybold, whose trays had been 75 mm high with partitions at 75 mm centres, patented an 'insert' which halved the space width (see Fig. 37), but later moved to an extruded section with 25 mm spacing, on which they standardised to keep costs down. As far as could be ascertained, the temperature gradient along the relatively thick fin was insufficient to make much difference, at the heats recommended, to medium sized (1 cm side) vegetable dice, either early on or towards the end of drying, but particle size was most significant. The residual frozen 'core', where it might be regarded as an entity, tended to take the form of an elongated vertical oval (rather than elliptical) shape as it receded. The last remnant seemed to be about one-third up, determined by interruption, on larger spacing trays, or on slab forms neatly filling the gap in narrower ones. Stokes favoured wider spacings, relying on simply the depth of loading per tray to arrive at an optimum, for a given material or particulation. There has been little investigation of the universality of particular tray configurations since then; having purchased a set, operators tend to adapt their usage to that design, if they depart from their original requirement.

HEATING SYSTEMS

Whatever the nature of the tray, dish or other device holding the material to be dried, the heat input chain will include multiples of some sort of shelf or horizontal plane area which may have a cooling feature but is principally designed to transmit the latent heat of sublimation towards the frozen substance by conduction or radiation, or a combination of both, at variable rates to suit the process. These layered structures (Fig. 38) may be heated electrically or temperature-controlled by circulating fluids. The common difficulty to be overcome is how best to spread the heat supply evenly.

Electrical circuits typically are formed by strapping bendable, metal-sheathed, mineral-insulated, heavy copper or aluminium rod (e.g. 2 mm diam.) in meander fashion to the underside of a metal shelf. This should be thick enough to even out the heat reaching the upper surface, and sturdy enough not to buckle. The spacing between shelves is such that the scatter or radiation effect is also fairly even, and indeed some arrangements position the product trays midway in the space, and not resting on a contact heating surface, so that advantage may be taken of the more penetrating effect of higher heater temperatures. Due consideration has to be given to the probable depth of material on the tray. Response is rapid, control very fine, and evenness of temperature over the whole array of shelves very regular, although of course (as always) this depends in the event on the rate at which heat is accepted as a sink by the material laid out to be dried.

Trials have been effectively conducted with a type of carbon-fibre woven mat, resin-embedded in a fibreglass shelf, called Mho-glas as a reference to the insulator classification of carbon, whose electrical resistivity *decreases* with rise in temperature. The network of the mat makes for very even heating, completely protected against oxidation, and with a thermal conductivity overall adequate for the purpose. The composite could be (and of course now is) moulded into other than that original double-faced flat sheet. This work at MAFF, Aberdeen, terminated before use of this construction material became general

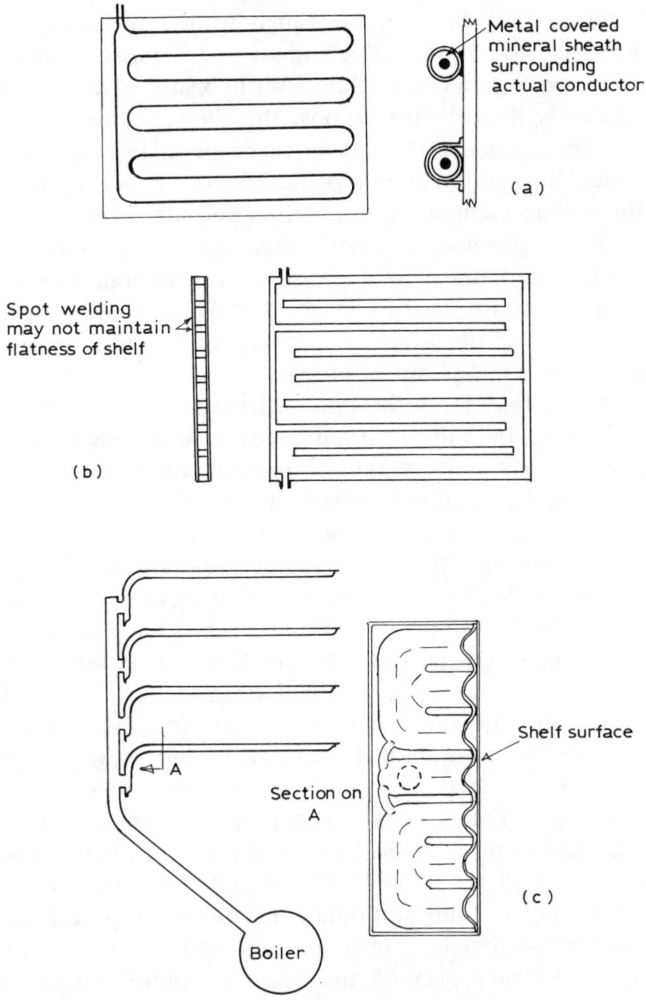

FIG. 38. Sample styles of shelf-heating means. (a) Solid/electrical, where evenness of supply heat depends on radiation and/or thickness of metal shelf. (Interleaving, freely suspended 'wire' grids are also used.) (b) Liquid, where the tack-welded labyrinth partitions act as stiffeners to the double skin. (c) Gaseous, where pressed (steel) undersheets stiffen the cantilever shelf and act as self-draining channels.

and its economic evaluation sufficiently favourable. A maximum demand figure for accelerated freeze-drying of raw cod steaks was 3·4 kW/m^2.

Fluids for hollow shelf heating range from refrigerants, of several types, through mineral oil, water and steam. External heat exchangers may require an intermediate fluid. With brine, glycol, ammonia or more commonly fluorohydrocarbons, the shelves may be cooled so as either to freeze produce directly in place or at least prevent it from thawing out during loading and pull-down. Freezing the produce within the drying cabinet by other than sub-atmospheric evaporative technique is thought not to be truly economical of chamber time.

The temperature range limits of some refrigerants impede performance also during the heating stage. Mineral oils allow of mild steel construction. Water has a better thermal conductivity than most, and generally requires less pumping effort.

The meander labyrinths through which such fluids pass within the shelf are usually made up from bars of rectangular section stitched or spot-welded to a relatively thin skin-plate, above and below. This makes for a strong platform which need only be supported at the (long) edges even if part of a movable array. The bars are arranged for fast flow, so that the pathways are sub-divided to give (say) four streams confluent after each traverse, and using convenient corners for flexible hose linkage to inlet/outlet manifolds. Quick-release couplings with self-sealing valves permit disconnection on occasion. In at least one system the shelves are assembled on a trolley, so that the trays with produce already frozen may be loaded on to it *en bloc,* the car then being wheeled into the chamber and the hoses only then coupled up.

Some designs of shelf have a dimpled or embossed fluid path pressed into the bottom sheet, the top sheet being flat and welded to it. It is rare to find a pathway fully looped, i.e. where the incoming fluid runs parallel to (and alongside) the outgoing same fluid through all its windings, so that the temperature drop (which overall can be 5° or more) across any part of the plate is minimised or averaged. Response through the skin-sheet to change of temperature is rapid compared with through the bars.

One system uses an arrangement of Γ-shaped hollow cantilever shelves, filled with steam at sub-atmospheric pressure(s) from a small heat-exchanger type boiler, so that saturation temperatures anywhere between 20°C and 100°C may be selected. This setting is maintained by the steam giving up its latent heat only and condensing, to run off into the down leg and be replaced by fresh steam, seeking to fill the void recurring where it is in demand.

Extruded sections in aluminium, having internal struts which become labyrinth walls given cut-away ports and end-sealing pieces, have been brought together with economy and precision to form very wide shelves similar to those in plate freezers.

In practical terms it would be a needless expense to start off with an optically flat metal shelf, with no unevenness caused by whatever method of construction.† Foodstuffs are not placed directly on its commercially flat surface but are manoeuvred by means of trays or dishes, very often of matt aluminium which, if heating is not entirely radiant, still makes general lightly weighed-down contact to good effect. The skin surfaces themselves spread and even out the heat flow, making up for any incidental inaccuracies. On the premise that the actual foodstuff-supporting contact will finally feed in the energy effectively, given the capacity to accept and ration out the heat transmitted by way of (or because of) the interspersions, which may vary due to production wear and tear, then the major element in the random chain of series-parallel connections is a high thermal conductivity in the metals. Witness the acceleration achieved by expanded metal inserts with a network of point contacts, a soft dry layer on one side and a physically resistive but imperfect metal surface to press against on the other.

MICROWAVE HEATING

Conduction or radiant heating during freeze-drying both depend on a high temperature difference to transfer the heat of further sublimation through the surface-origined dry layers, whereas high-frequency energy can pass through such layers to be preferentially absorbed at the ice front. Foodstuffs considered in terms of dielectric properties are complex indeed. The alternating currents enticing their polar molecules to align at many megacycles per second, so creating the

† An early consideration was the possible use of a new electric heating material called Mho-glas, a mat of carbon fibres embedded in glass fibre and resin, which could be moulded into a shelf/tray combination. Made by a company of the Hunting Aircraft Group in 1960, it soon found uses elsewhere.

internal heat, are in the radio-frequency range and governed by country law. The two most prevalent (if not most popular) settings for microwave heating energy are 915 MHz (896 in the UK) and 2450 MHz (2375 in the USSR); this to some extent defines their penetrative power.

The cooking, tempering and thawing, and even drying uses in normal atmosphere of microwaves, are in widespread domestic usage today, as well as in industrial experience, but in vacuum processing there has been some limitation because of possible corona or glow discharge. Consequently, for some time the absolute pressure range much below 50 mm Hg (6·67 Pa) was avoided except in research work, although it was quite clear that, provided the input energy was matched properly to the rate at which water vapour could be produced, then the energy surplus which showed up as ionisation of the air in the chamber would not be there.

The optimum pressure condition for that occurrence is 1 mm Hg (133 Pa), a typical operating pressure for many freeze-drying operations. Some workers have sought to avoid the phenomenon by very low cabinet pressure, involving 5-stage steam ejectors and 1000 MHz. Others, using 2450 MHz and refrigerated condensers, reached 0·1 mm Hg [even 0·05 mm (6·65 Pa)] as a scavenge to avoid the problem; since 0·05 mm corresponds to −46°C, and the condenser temperature may then be −52°C, this is a costly solution for avoiding voltage breakdown.

As freeze-drying progresses, the power must be reduced to match the load, leaving no large excess to support flashover. Much can be done during design and construction stages, in avoiding rough edges or metal projections which might trigger arcing, although some defects in appropriate detail can be masked by aluminium tape. High field strength across (Teflon, better fused silica) waveguide windows has been a troublesome problem, where microwave power has been delivered from waveguide penetrations on opposite walls to the space between resistance heated platens. A simpler approach might be to feed microwave power to the load from antennas placed between the platens and above the load, so that energy could radiate directly down on the product. The normal atmospheric breakdown strength of air, around 30 000 V/cm, may be less than 100 V/cm *in vacuo*.

Dielectric property differences have been detected between the aqueous matrices intra- and inter-cellular, and the more complex sub-unit systems adhering to cell membranes and walls, which latter may not have frozen out, and may therefore excite preferentially. Models for energy coupling by geometry and size of product, contrasting its intrinsic impedance to that of free space in the cavity, while assuming attenuation factors, have led to forecasts of efficiency, but improvement in estimates of reduction in terminal drying time is still empirical. Perhaps owing to shorter processing times, however, clear advantage is reported in retention of volatile components.

The US Army Natick R & D Command (Anon., 1984) found microwave dispersion of residual moisture (some 20%) useful in foodstuffs, from entrées to fruits, which had been partially freeze-dried to a stage where their bulk could then be reduced by compression. Reconstitution included a springy reversal of the compaction in volume.

It is interesting to note that the Natick workers (Decareau, 1985) on microwave freeze-drying observed that tiny ice crystals could detach and dislodge from inside the parent foodstuff, and be transported to the ice condenser by the general water vapour mass transfer 'whirlwind', without apparently having had to give up their latent heat!

Summarising, no technical impediment remains to prevent microwave freeze-drying of foodstuffs from taking off. Ideally there should be continuous power-to-load control, with sensing of the onset of corona, compensated by reducing the power automatically. A comprehensive review of microwave freeze-drying is given by Sunderland (1982).

HEAT PUMPING

The idea of heat pumps stems from Kelvin's warming engine concepts, based on fluid cycles where advantage is taken of change of state latent heat at (say) two levels of temperature.

For the most part refrigeration machines function for the prime purpose of heat absorption at the colder level, with considerable heat rejection at the warmer level to a nondescript sink. A so-called *heat* pump, on the other hand, is a refrigeration machine whose purpose is looked on as *supplying* heat at the higher level while absorbing heat (from a nondescript source) at the lower level. Neither sink nor source requires to be inexhaustible but of such entropy that during the process their temperature level remains virtually unchanged.

In freeze-drying, however, there is opportunity for *useful* transfer at both high and low temperature levels, since heat of sublimation is needed equally as well as withdrawal of heat for ice-vapour condensing. Both are within the envelope of the process. Such 'chicken and egg' dependence seems to embarrass.

In laboratory driers, compact machines with user-friendly stop/start switching, arrangements where the refrigerator is first employed to freeze the sample, then to furnish the ice condenser, have been popular for some time. The heat of sublimation in such a case is often supplied by the ambient air, which might also be called on to aid defrosting. But on somewhat larger (still intermediate size) pilot plants a secondary fluid tends to be incorporated, with independent heating and rapid cooling means, in the name of flexibility rather than call for dexterity in manipulating the actual refrigerator. Such thinking tends to follow through on industrial plants. Yet in multi-cylinder machines, for instance, more than one off-loading pressure can be arranged, and hence a truly balanced if intricate response, to all aspects of the process, can be devised. One has to set aside the view that the trapping of a vapour as ice is a consequence of vaporisation from ice; as far as the continuum of the process is concerned they are complementary. The driving machine has to have capability of adjustment between the two, but overall efficiency of performance will still be well in hand.

FLUID PUMPS

Mechanical pumps for gaseous substances gulp regular but equal *volumes* rather than mass. The amount of 'stuff' they take in at each revolution or intake stroke is limited by the amount of gas residual from the previous working cycle, which in turn depends on the instantaneous conditions. For freeze-drying, such pumps (Figs 39 and 40) have two primary functions:

(1) To establish control over the heat supply to the material being dried from the frozen state, by removing conductive non-condensable gases within the drier or its locks; these may be (initially) at room temperature and pressure, and continue to be pumped until their presence is of little or no consideration.
(2) To maintain that arbitrary condition in the face of subsequent seepage, inadvertent or deliberate.

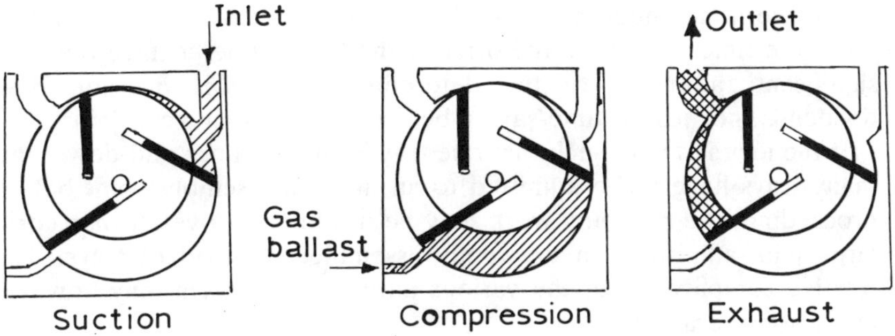

FIG. 39. Working cycle of a rotary vane pump.

Their third but unofficial requirement is the practical one of accepting and dealing with unavoidable vagrant water vapour, which may present itself suddenly in an appreciable amount or more likely in stray wisps.

It will be seen that during pump-down the mass rate of gas dealt with at constant cycle speed will diminish logarithmically, but the volumetric efficiency of the device—actual vapour volume handled versus ideal (no clearance or slippage) volume—will remain virtually constant.

The pump-down time to start-up is sometimes determined in plants with simple instrumentation, by a level of non-condensables partial pressure which is low enough to permit the total pressure read to indicate that the water vapour partial pressure is itself below the triple point, i.e. 4·58 mm Hg. For the freeze-drying of *foodstuffs* as a class, contrasted with living biological materials or pharmaceutical substances, this total pressure start-up signal is commonly 1 mm Hg, of

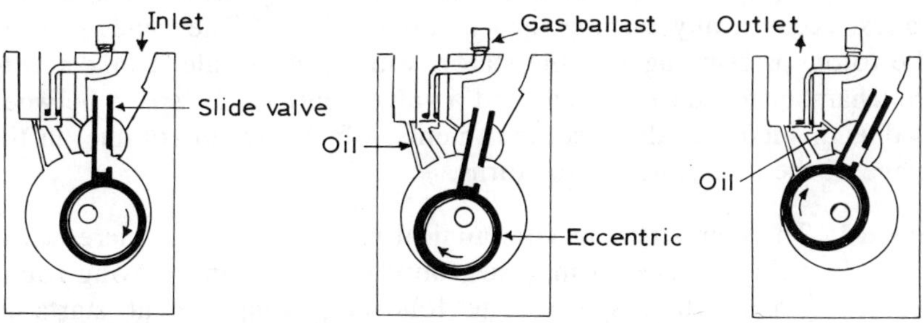

FIG. 40. Working cycle of a rotary piston pump.

which the non-condensables partial pressure may well be 0·1 mm Hg. Since the time to deprive the drier of the last few percentage points is significant, attention to the determination of this decision pays dividends, not just in time-saving but in equipment choice.

If the lock is small, such that one tray is admitted per pull-down, but many trays have to be admitted to an internal assembly point before proceeding, the pumping work may be no different overall than for a large lock admitting an externally assembled car-load of trays. The relative complexities of the various parts of the system may however invite consideration.

Sub-atmospheric pressure work is referred to (PNEUROP) as 'rough' vacuum between 1000 mbar and 1 mbar and 'medium' vacuum from there down to 10^{-3} mbar; 'high' vacuum (to 10^{-6} mbar) and 'ultra-high' vacuum (below 10^{-7} mbar) need not concern us in commercially viable food freeze-drying, and indeed the process is looked on as just into the 'medium vacuum' range, so that relatively the gaseous pumping facility it requires and is serviced by is of generous efficiency and quality.

Ring, i.e. centrifugal, 'fan type' pumps are able to handle 'wet' vapours from about 54 mbar (40 mm Hg) upwards; rotary vane pumps with gas ballasting can be used confortably down to 1 mbar; more generally in large industrial foodstuff freeze-drying installations a 'dry' gear or 'blower' pump is used for pull-down and continuous operation, backed by a liquid-ring pump. Sufficiently fast times can be achieved without resorting to two-stage Roots pumps followed by two-stage backing pumps, as in metallising etc. where the process time relative to the set-up time is so much shorter.

Rotary Vane Pumps

This type (Fig. 39) comprises a rotor equipped with three sliding vanes, eccentrically turning in a (stator) housing. The vanes slide in the rotor, pressing against the stator, which has an inlet connected to the chamber to be evacuated, and a valved outlet. The space between is thus divided into three compartments, which vary in volume, in the plane of the pump, as a rotor turns.

Stage 1. The compartment communicating with the inlet increases in volume, sucking in gases, until the next vane moving round isolates that space. The following compartment starts its cycle.

Stage 2. The isolated compartment becomes smaller, compressing the gases to a higher pressure.

Stage 3. The advance vane of this compartment uncovers the outlet port, while the compartment continues to get smaller. The pressure continues to rise, until it reaches atmospheric pressure, when the exhaust valve opens, and the gas(es) can be pushed out.

To ensure a seal between vane(s) and stator, *and* on the *side* walls of the pump, lubricating oil is present; some of this is carried towards exhaust with the gases. Large oil droplets will fall into the sump, but fine droplets have to be caught by a filter, becoming transformed into large enough droplets to be re-injected into the body of the pump. Clean gases and vapours are thereby exhausted.

Should the inlet draw in *condensable* vapours, however, the conditions within the pump must be such that their saturation pressure(s) at the prevailing temperature is not obtained in Stage 2. Then there would be *no* further compression, but condensation. The exhaust valve would be unable to lift. The liquid phase (usually water) produced would mix with the oil and contaminate it, forming an emulsion, which would upset its lubricating function, clog the filter and generally damage the performance of the pump. It is possible to meet this problem in several ways, but the standard method is to 'gas ballast'.

The pump is designed to compress from 1 mbar to 1000 mbar, a compression ratio of 1000:1. By arranging a (gas ballast) valve to inject a given quantity of atmospheric air into Stage 2, thereby raising the pressure there 'artificially', the pump has only to perform a ratio of (say) 10:1 to be able to eject to atmosphere. Thus its previous inability to pump or get rid of inherent water *vapour* will be overcome. But there is still the question of the liquid phase. This hinges on the normal running temperature of the pump and the corresponding saturation pressure of water vapour. Pumps should always be allowed to attain their running temperature before being put on duty.

A typical 3-litre vane pump *on its own* will lift (say) 0·002 25 g vapour from 1 mbar (saturation temperature −20°C) to 1000 mbar in 1/25 of a second, in a machine mass of perhaps 180 kg held at around 70°C (saturation pressure 317 mbar) by a cooling fan. On being compressed to above the triple point, the vapour is able to liquify, *provided* its total heat content can adjust appropriately against the work being done on it in this environment. Clearly this possibility cannot be left to chance. Pump makers stipulate perhaps 30 times this amount of water vapour at no more than 30 mbar pressure with the gas ballast valve in operation, but then the end (attainable) pressure may be limited to 4 mbar rather than 1 mbar.

The protective function of the refrigerated ice-vapour-condenser over the vane pump's contrived ability to handle a limited amount of water vapour may be strained during evaporative freezing, or when the ice-condenser is already well loaded but just prior to defrosting, or inadvertently because of malfunction. But ordinarily such pumps, particularly on laboratory, pilot or medium-sized industrial cabinet freeze-driers, perform admirably in arduous conditions with regrettably sometimes scant attention.

Roots Blowers

This positive displacement pump type (Fig. 41) has a wide range of application in vacuum fields where high pumping speeds at low pressures are sought. In series with other pumps these dry-running compressors can do duty on both condensables and non-condensables, ranging from high-vacuum work to above-atmospheric pressures. They have been used to propel vapour towards an *intermediately placed* refrigerated condenser followed by an oil-sealed rotary pump, but more commonly for freeze-drying they are positioned between the ice-condenser and a backing pump.

The two lobed symmetrical rotors have a figure-of-eight cross-section and intermesh to rotate in opposite directions, being interlocked by a gear drive. The synchronised rotors move past each other and the housing with very small clearances, making for high rotational

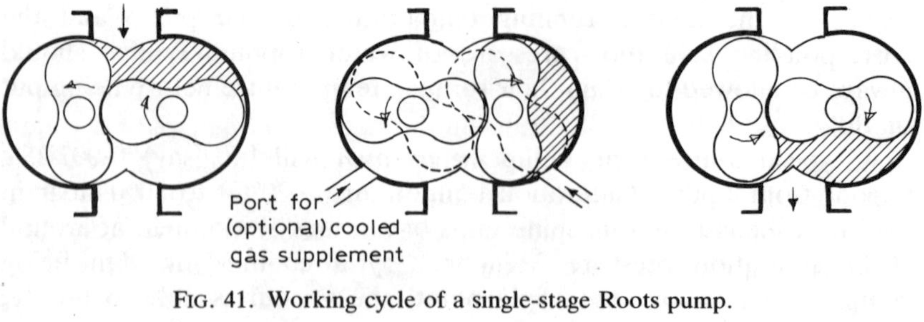

FIG. 41. Working cycle of a single-stage Roots pump.

speed and therefore a high (volume flow) pumping rate. The dynamic nature of these 'seals' sets limits on the compression ratio possible, so the *differential* pressure gain achieved normally runs up to 53 mbar or 160 mbar. From then on, the pumped gas load is compressed to atmospheric pressure by an oil-sealed rotary vane or water ring pump, for instance.

The gearbox and bearings of Roots vacuum pumps are oil-lubricated, that chamber being evacuated through piston-ring seals on the impeller shafts. Sometimes the rotor of the drive motor is direct on the pump shaft, in the vacuum space, the stator being outside a vacuum-tight 'can' of non-magnetic steel.

A built-in pressure compensation valve and integrated bypass line allows a Roots pump to be started at atmospheric pressure together with the backing pump, without overload in the higher pressure range, even where the ratio in pumping speeds between the pumps is high. Gain in pump-down speed with the valve may be as much as 30%.

Compatibility with the follow-on pump is important. As the Roots pump chamber opens to the outlet port, a warm stream of higher pressure gas may enter, compressing the displaced gas volume. So, just beforehand, a supplement of precooled gas is arranged to enter the chamber, precompressing the pumped gas. The heat of compression is thus largely eliminated. In multi-stage pumping systems this cooled gas is drawn from series-connected gas coolers; in single stage systems it is from the atmosphere.

In a multi-cabinet freeze-drying installation, for instance, a large two-stage Roots pump backed by a liquid-ring pump may be used centrally and sequentially for fast pull-down, handing over to a smaller, similar series of 'holding' pumps individual to each cabinet.

REFRIGERATION DEVICES

An elementary circuit for a refrigeration plant is indicated in Fig. 42. Heat is transferred from an environment at low temperature to another at a higher one by utilising the natural properties of a volatile fluid during vaporisation (for which it demands or absorbs heat) and condensation (when it gives up or dissipates heat). In the vapour-compression type a mechanical compressor moves the refrigerant fluid, from a condition of lower than its saturation pressure at the lower temperature, to a condition of higher than its corresponding saturation

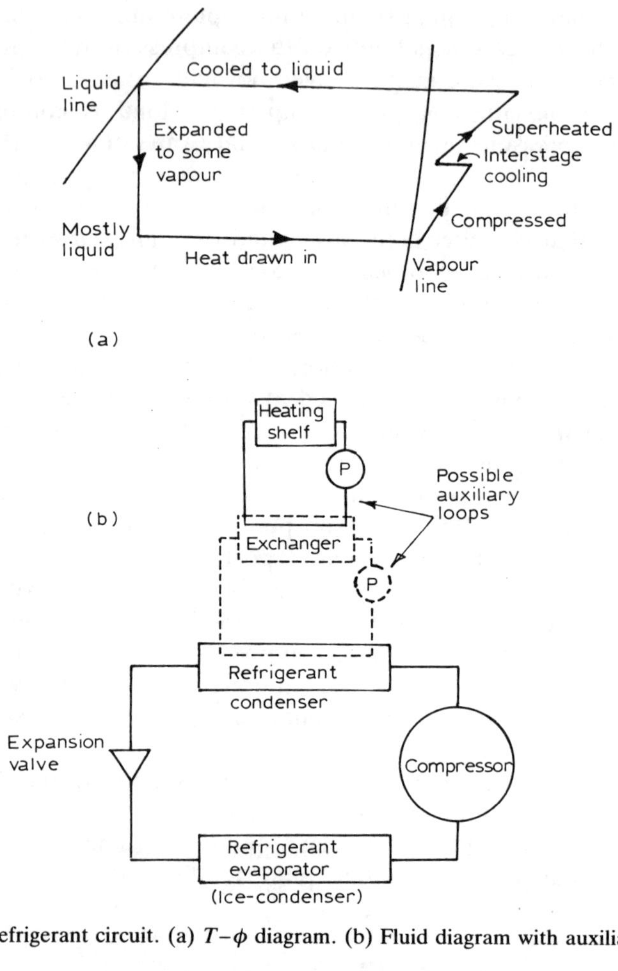

FIG. 42. Refrigerant circuit. (a) $T-\phi$ diagram. (b) Fluid diagram with auxiliary loop.

pressure at the higher temperature, while in the absorption system type the transfer is effected by the input of heat from a 'boiler'.

The evaporation of the refrigerant, i.e. the heat-demanding part of the cycle, is performed in a usually tubular, sometimes of hollow-plate form, so-called evaporator, which from the point of view of the freeze-drying process itself is termed the *ice*-condenser. In giving up its latent heat of 'vapour to solid' phase change, to the refrigerating fluid, thus enabling *it* to become a vapour, the water vapour from the subliming product again changes phase, 'condensing' as snow or ice on the deliberately arranged cold enough surfaces comprising that part of the circuit.

The condensing of the refrigerant is at the higher temperature level, and is therefore capable of being carried out by means of abundant cooling water, for example. This function would be aptly carried out in a (refrigerant) condenser but this essential part of the cycle tends to be glossed over, undoubtedly because of possible confusion in nomenclature. The exposure of this (water) coolant in an auxiliary spray, cascade or 'evaporative chiller' section, with some loss and make-up, can further complicate the word-picture, but the chain of fluid loops is linked together by phase changes (Fig. 42(b)). Viewed in this manner it has even been arranged so that heat from condensing the refrigerant is used to supply the heat of sublimation of the product ice, in a 'heat pump' concept.

The work requirement would be reduced if practical compression were carried out nearer the ideal of constant temperature rather than at the other ideal of constant heat content, so multi-stage compression with intermediate cooling of the superheated vapour is employed to bring the vapour condition periodically towards the isothermal line.

The efficiency of a refrigerating cycle is measured against a 'coefficient of performance'—the ratio of the heat removed by or *to* the refrigerant to the heat deliberately added—thus

$$\mathrm{COP} = \frac{\text{Refrigerating effect}}{\text{Work done on refrigerant}}$$

$$= \frac{\text{Heat absorbed in evaporator}}{\text{Heat added (say in compression)}}$$

$$= \frac{\text{Heat removed in condensing ice}}{\text{(say) Heat put in at generator}}$$

The 'theoretical (inverse Carnot) maximum efficiency' is given by

$$T_{\text{high}}/(T_{\text{high}} - T_{\text{low}})$$

For absorption plants this translates to

$$\frac{(T_{\text{cond}} \times T_{\text{gen}}) - (T_{\text{evap}} \times T_{\text{abs}})}{T_{\text{cond}}(T_{\text{cond}} - T_{\text{evap}})}$$

and in practice is unattainable, because of various internal inefficiencies.

The standard ratio, against which other performances can be gauged, is 4·6 for ammonia, when operating at an evaporating temperature of −15°C, condensing at 30°C, the liquid before the expansion valve being at 25°C and the suction gas at 10°C.

A machine under these conditions would require, 'per ton of refrigeration', NH_3 refrigerant of amount (kg/min)

$$\frac{211\ \text{kJ/min}}{\text{Heat in 1 kg leaving evaporator} - \text{Heat in 1 kg entering expansion valve}}$$

the denominator being obtained from enthalpy tables, given the appropriate temperatures as above.

This 'nominal ton' of refrigeration refers to a ton of ice melting in 24 hours, standardised to a heat rate of 1 TR = 200 BThU/min. Normally one would not expect an absorption plant to be installed on a duty of less than 100 TR = 300 000 kcal/h.

In freeze-drying, the evaporator surface area generally is fixed, and the compressor (or boiler) settings are altered to suit a varying load. The cooling water supply may have to be regularly adjusted. Too high a refrigerant temperature at entry to the condenser will indicate wasteful operation. The suction temperature usually has to be lowered as a drying cycle progresses, or kept low on continuous operation, giving a relatively larger specific volume on entry to the compressor. If the volume rate handled drops, this means a lesser refrigerating effect. On a batch plant this may not matter, as the ice deposition rate will also be falling off. Should the snow/ice layer(s) be such as to raise the point of deposition temperature unduly, however, then the refrigerant temperature must be caused to fall, or the ice-layer effect countered by defrosting.

The vapour-compression refrigeration cycle in essence is a reversed Rankine cycle which differs from the reverse Carnot cycle in two particulars: (a) it makes use of an expansion or metering valve to give an irreversible adiabatic expansion which is slightly different from the vertical isentropic expansion line in the $T-\phi$ diagram, and (b) it uses dry compression instead of the theoretical wet compression, exceeding the (ambient) temperature irreversibly to a state of superheat, thus reducing the coefficient of performance below the idealistic 6·6 (the inverse of 15%).

The tailing of duty *amount* in a batch cycle leads to off-loading in multi-cylinder reciprocating compressors, by discharge bypass for instance. The flexibility of the various rotary machines with interruption or pauses is also resorted to.

The throttled expansion of liquid refrigerant, from a high pressure in the (refrigerant) condenser to the lower (perhaps 7:1) pressure in the evaporator, is done in driblets, not only to evaporate partially, thus cooling the remainder but, by providing (say) a fixed liquid level, the 'flooded' control can be manipulated. Thus the lowering of evaporator temperature to cater for end-effects, such as thicker ice-formation, is achieved. 'Continuous' machines have a more elaborate control linking system, especially if they have several evaporators per compressor.

The absorption cycle accomplishes compression by using a secondary fluid (e.g. water) to absorb the refrigerant (for water its companion would be ammonia) when it is a gas. The principle behind it is that the vapour pressure of a refrigerant will be lowered by the addition of an absorbent which has a lower vapour pressure. The greater the quantity of absorbent used, the greater the depression from the vapour pressure of the refrigerant. There are thus two interlinked cycles, that of the refrigerant and that of the absorbent (see Fig. 43).

The condenser, receiver, 'expansion' valve and evaporator are virtually the same as in any vapour compression cycle. The compres-

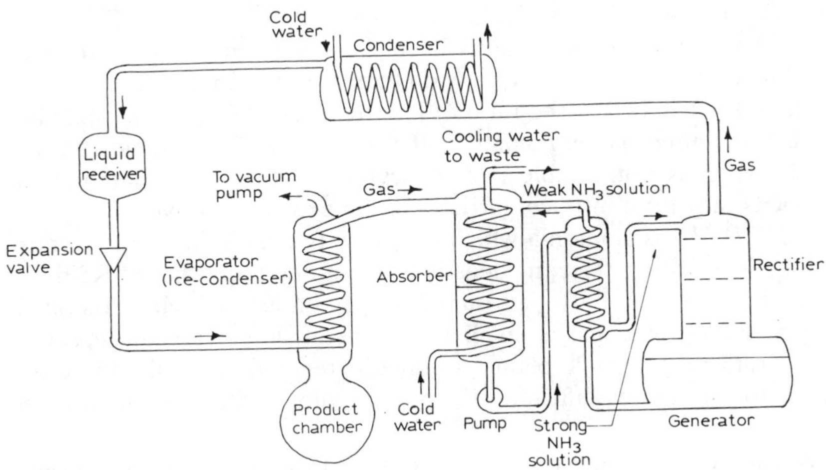

FIG. 43. Absorption cycle.

sor, however, is replaced by an absorber for the refrigerant leaving the evaporator at low temperature and pressure, a generator where heat is applied to distil the refrigerant at high temperature and pressure, a pump because of the difference in pressures, a heat exchanger, and a controlling pressure reducing valve (say) between the heat exchanger and the absorber. Thus the vapour pressure of the solution in the absorber, maintained at suitable temperature and concentration, can be kept lower than that of the refrigerant in the evaporator.

Spraying the weak solution in the absorber then causes refrigerant vapour to flow from the evaporator to the absorber. The strong solution thus formed is pumped via a heat exchanger to the boiler or generator where steam heat or direct means releases the refrigerant vapour. A highly efficient rectifying column ensures clean gas. The remainder of the refrigerant circuit is the same as that in Fig. 42.

In general, refrigeration circuits of the kind associated with freeze-drying plants can endure cooling water supply temperatures higher than those required for steam ejector systems, and absorber systems are easier to adjust to a varying load than mechanical compression systems. Refrigerants for industrial size plants are mainly R502 and R717 (Table 15), with R12† and R22 receding in popularity, partly because the EEC has frowned on upper-atmosphere contamination by halocarbon aerosols, but also because insolubility with oil is an important practical feature.

R502 is an azeotropic mixture of R22, monochlorodifluoromethane ($CHClF_2$), at 48·8%, with R115, chloropentafluoroethane ($CClF_2CF_3$), at 51·2%. The two components have the same concentration in both liquid and vapour stages over a wide temperature range, acting as one, so that if a leak occurs the ratio of loss remains equal. The permissible moisture content is very small, half that for R22 but double that for R12. Water is not soluble in fluorocarbon gases, and isolated small droplets may form into ice particles in the expansion valve.

R717 (NH_3) is ammonia which of course is soluble in water. It must not be used with copper; its density is about one-tenth that of R502 so, while it is easier to move, its pumping speed must be higher. Its much higher latent heat makes small variations in flow a greater upset to temperature; on small plants automatic regulation would therefore tend to be more difficult. Even so, with R502 the hunting in

† Recent research by teams at McMurdo Sound has connected the autumnal hole in the ozone layer, above Antarctica, with CFCs.

TABLE 15
Comparative Properties

	R502	*R717*
Molecular mass	111·63	17·03
Boiling point at normal atm. (1·013 bar) (°C)	−45·4	−13·3
Latent heat at b.p. (kJ/kg.mol)	19 258	23 343
Evaporator pressure at −15°C (bar abs.)	3·49	2·35
Condensing pressure at +30°C (bar abs.)	13·19	11·64
Compressor discharge temp. (°C)	37–55	99–127
kW per ton at −15/+30°C	0·805	0·738
at −23/+38°C	1·58	1·31

Data from Anon. (1981).

ice-temperature which can occur in keeping with the opening and shutting down of the expansion valve on cabinet plants may amount to several degrees.

All the properties of a refrigerant which would render it ideal or at least truly desirable for a freeze-drying operation on an industrial scale are difficult to find in one substance (Table 16).

The efficiency of the cycle:

$$\frac{\text{Condenser temperature}}{\text{Difference between condenser and evaporator temperatures}}$$

is related to the corresponding pressures. Consequently, a compressor suction intake which is above atmospheric pressure, and a discharge pressure which is around 14 bar (1380 kPa) so that the compression ratio is kept down, will be preferred, and make for a better volumetric efficiency.

The latent heat per unit specific volume should be large at the evaporator, to decrease compressor work and size. Sensible heat content is circulating; capture by sub-cooling prior to the evaporator will reduce adverse throttling effects, so the latent heat proportion should be large and the vapour volume compact, making for a smaller volumetric capacity, or a slower compressor speed. The oil returning ability at evaporator suction is also important.

It will be clear that off-loading a mechanical compressor, which must incur some wasteful work, may be relatively more complicated than regulating the 'stoking' of an absorption machine, size for size.

Refrigerant condensers (Figs 42 and 43) are of the shell and tube

TABLE 16
Properties of Water/Ammonia Solutions (Intrapolated)

Condenser pressure (atm)	13·6	13·6	11·6
Condenser temperature (°C)	+36	+36	+30
Suction pressure (atm)	1·02	2·04	3·06
Suction temperature (°C)	−34	−18	−12
Strong solution (% wt water)	20	30	40
Vapour (% wt ammonia)	99·99	99·97	99·90
Temperature (°C)	−29·1	−7·4	+10·8
Liquid heat content, h_f (kJ/kg)	−271·6	−230	−188·4
Enthalpy of vapour, h_v (kJ/kg)	1229	1268	1303
Weak solution (% wt water)	8·6	20	31·5
Vapour (% wt ammonia)	100·0	99·99	99·95
Temperature (°C)	−32·2	−12·6	+3·95
Liquid heat content, h_f (kJ/kg)	−210·3	−197·9	−185·1
Enthalpy of vapour, h_v (kJ/kg)	1223	1255	1286

Data from Anon. (1981).

type, and of regular design. Vapour evaporators as ice-condensers, on the other hand, are various; equipment manufacturers differ in approach (Fig. 44). They all of course endeavour to avoid vapour locks at coil bends, and trapping of entrained oil which might reduce heat transfer capacity as well as depriving the crank-case (Figs 45 and 46).

All endeavour to mask the draw-off point of the non-condensable vapours. Some mask the route to the condensing surface. Concepts of

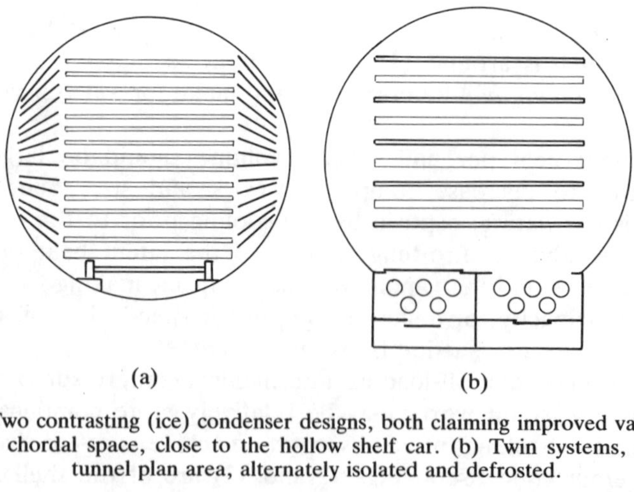

(a) (b)

FIG. 44. Two contrasting (ice) condenser designs, both claiming improved vapour flow. (a) In the chordal space, close to the hollow shelf car. (b) Twin systems, within the tunnel plan area, alternately isolated and defrosted.

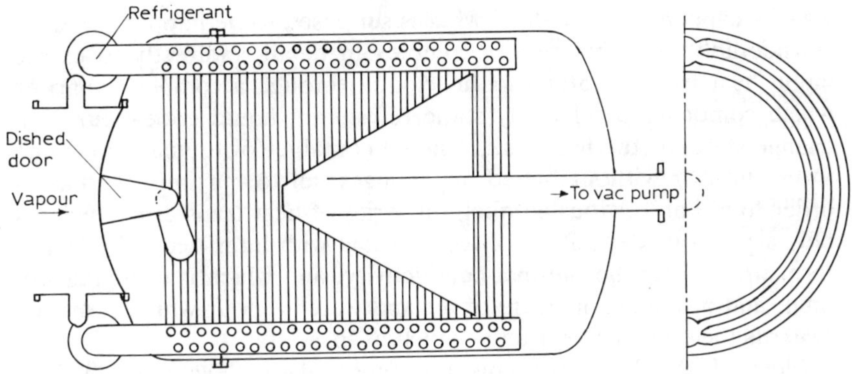

FIG. 45. Ice condenser, style A; note protection for vacuum pump suction.

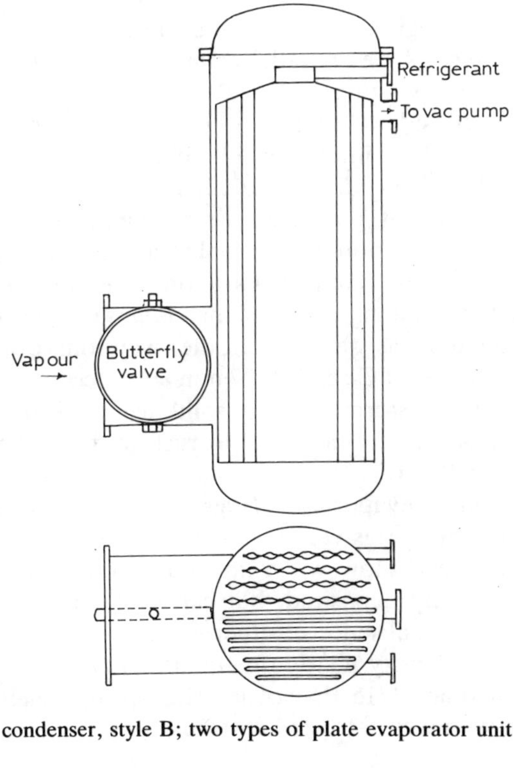

FIG. 46. Ice condenser, style B; two types of plate evaporator unit are indicated.

what is happening, or indeed what is supposed to be happening, differ considerably. The function is complex in detail, indirectly to extract latent heat from a water vapour 'cloud', through a composite barrier, which comprises a relatively static refrigerant liquid endeavouring to change state, as the heat sink, a metal container of tubular or channel form, with or without finned appendages, all to maximise surface in order to aid an obviously unhelpful series of heat transfer coefficients, plus a previous deposition of ice *or snow* with interstices of irregular form which alter the thermal conductivity, and which is changing with time, and a tenuous mixture of air and water vapour which is moving towards the condensing area as much as 'past' it.

Most of the larger designs are reported as developing bulbous 'noses' of ice-formation on the part(s) of the evaporator directly confronting the approaching vapour, leaving the rear portions bare, even when these parts seem to be fed just as evenly with refrigerant. Calculated dimensioning of the configurations has to make assumptions concerning the spaces *between* the component parts of the evaporator, which ought to prevent this phenomenon, taking into account, for instance, series/parallel arrangement of the 'sub-circuits' and tapering-off of volumes uncondensed. But it still happens. Admittedly little consideration can be given to shielding the vapour and leakage air from heat interchange within, or with the walls of, the vacuum spaces. The foodstuff ice-vapour is undoubtedly superheated thereby in the course of practical freeze-drying, and therefore less ready to condense; the non-condensables are not generally heated enough on arrival at *their* pump. Separation of the two streams has to be achieved by the pumping effect of the *collapse* of the vapour. This can pump entrained air as well as the other way round. Water vapour of amount 1 g, at −20°C and 0·77 mm Hg, occupies a space of 1·133 m^3, and on transformation to solid ice it shrinks to $1{\cdot}09 \times 10^{-6}$ m^3 when further molecules have to rush in to fill the 'void'. The contraction is over $10^6:1$.

Finning on tubular evaporators (Figs 47 and 48), reminiscent of high-velocity-air cooling rigs where air-flow is preponderant to vapour flow, tends to produce snow when applied to rarefied air work; the snow may be unevenly deposited, but can be tampened. A clear-ice deposition indicates an absence of other gases.

Snow has a bulk density nearly one-quarter that of ice. Its thermal conductivity is perhaps 1/15 that of ice. It appears beneficial to start off refrigerating with wet defrosted surfaces.

FIG. 47. Interior of ice condenser showing protection for vacuum pump suction.

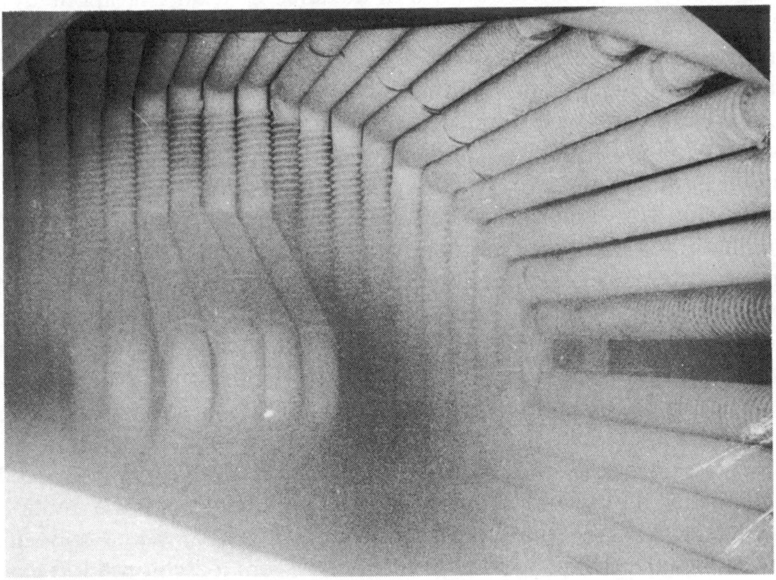

FIG. 48. Same interior as in Fig. 47 at the end of the run; note 'snow' distribution.

Horizontal labyrinth plates in the chordal spaces of a cylindrical cabinet tend not completely to dry off after wash water has disposed of the deposition. Vertical such plates have been sported with dimples as locking-on means to prevent ice falling off due to its own weight. Alternate use of twin ice-condensers limits the build-up of ice beyond a condition where its thickness presents an appreciable temperature (and therefore partial pressure) difference. How often this has to be done will depend not only on the thickness locally but on the evenness of coverage by non-condensable gases. At least one design, while taking care to guard against excessive vapour entry to the vacuum pump by a shielding sub-circuit, displaces by far the greater part of the deposition surface at radial, i.e. right, angles to the directional path for the interfering gases.

On the universal premise that Nature abhors a vacuum, and will take determined steps to fill one, plus the inescapable fact that she deals in partial pressures, it follows that the space beyond the refrigerated ice condenser 'coils' will still contain water vapour molecules to the extent that the temperature and therefore the corresponding saturation pressure or molecular activity around the 'trap' will allow. Needless to say, the non-condensables pumping system which is sealed with oil has in advance of it an oil-vapour-free void, as it were, and will tend to leak its oil molecules into the chamber if conditions are allowed to be such as to permit it. Choice of oil is therefore significant; it must have a low evaporation temperature, or rather its free surface molecular escape activity must not amount to much under the prevailing conditions (a typical vacuum pump oil at 80°C will have a vapour pressure of (say) 1×10^{-4} mbar = 0·000 075 mm Hg), and it must be capable of accepting some degree of contamination from water vapour without rendering the compression efforts of the pump ineffective. Recent suggestions have included superheating the water vapour to allow it to pass through the pump without emulsifying the oil, and of course new oils. Fortunately, in freeze-drying for foodstuffs on a factory scale, the conditions prevailing are not such as to present either of these difficulties as more than a token hazard.

It is some time now since an ozone depletion hypothesis was mooted and published (Rowland & Molina, 1974) arising from the million tonne release to the atmosphere (from aerosols, foams and solvents rather than refrigerant purposes) of fully halogenated chlorofluorocarbons (CFCs) whose chemical stability in the troposphere and/or stratosphere since their development in the 1930s had hardly been put to question.

The United Nations environment programme led in 1987 to a 24-nation adoption of a CFC protocol, developing into international control measures on their production and use. Of those existing and now regulated, which are used significantly in industry *as refrigerants* (therefore having some bearing on freeze-drying) *and* have a high ozone depletion potential (ODP), the order of demerit is: R11, R12, R114; R113, R500, R115.

Refrigerant R502 is in sight as being made up of R22 (48·8%) (unregulated) and R115 (51·2%) (listed). R22 has one-third the ODP of R11. In 1986, use in the UK general refrigeration industry of R11 (6·5%), R12 (34%), R22 (43·5%) and R502 (16%) showed that industrial use was insignificant compared with air-conditioning (38·5%) and commercial use (48%) while manoeuvrability was better. Proportionality of ammonia use in central machine houses of food processing factories is not clear.

The prospect of vapour compression refrigeration without most tried and tested CFCs is not cheering to commercial outlets, for instance, where spillage is rare, tending to be confined to scheduled overhauls and discarded units. The CFCs were developed by selection for purpose from the whole range of low moleculars. Higher molecular weight substances are not so suitable generally because of lower vapour pressures and higher freezing points.

The scrutiny of organic fluorine compounds having only one carbon atom, but containing no chlorine, or else at least containing chlorine in combination with hydrogen, which are not flammable and have low toxicity, results in only R22 ($CHClF_2$) being viable as an alternative for long-term acceptance. It may be objected to on 'greenhouse effect' grounds.

Similarly, looking for two-carbon fluorine compounds which might prove future alternatives, isomer R134a (replacing R12) is noteworthy, possibly R123 (for R11), with R124 and R125 not ruled out at the present stage. R134a is insoluble in mineral oils and requires a synthetic lubricant which is miscible at both low and higher temperature. It is suitable for all the many links in the freezer chain: centrifugal, reciprocating or screw compressors, such as are encountered in chilling, freezing, cold storage, transport and industrial systems.

Novel alternatives for R12 are currently estimated to cost five times as much and availability will depend mainly on toxicity clearance tests, which may take some years.

Change of refrigerant infers change or departure from specific attributes or dependent properties, which may entail modification of existing plant. Compression indices increase from R502 through R12 and R22 to ammonia. Capacities of evaporators and condensers are must less affected than compressor size. (Extraction of water vapour by steam jet does not, of course, require CFCs at all.) Practical aspects of implementation of the protocol on both existing and future plant design were considered at a 1988 conference in London (Ackroyd, 1988).

Chapter 9

Instrumentation

Pressure measurement instruments for installations that are subject to vacuum range from the approximate, quick indication types to the highly sophisticated, complex and exceedingly accurate (Fig. 49). Modern technology can now couple remote sensors of at-a-glance indicators of both pressure and temperature to control production systems on programmed cycles. Except perhaps for laboratory-scale investigation, freeze-drying operations need only take heed or be concerned with conditions above 0·01 mbar, so a number of devices catering preferentially for various ranges between 10^{-2} and 10^{-12} mbar will not be considered here.

The ubiquitous Bourdon tube gauge, although dependent on ambient atmospheric pressure, nevertheless offers clear linear indication on vacuum lines from 1020 to 'zero' mbar. A typical dial scale might have a division width covering (the last or any) 20 mbar of $3\frac{1}{2}$ mm. The heart of the gauge is a bronze tube of nearly elliptical cross-section, one end secured in a block connected to the vacuum line. The tube has been bent usually into a C-shape like a curved cantilever, the 'free' closed end being linked to a quadrant and spur wheel operating a pointer. Change in pressure causes the cross-section and therefore the curvature of the tube to alter slightly and, thus magnified, the position of the pointer on the dial. Its stamped-out gearing is simple; the gauge can withstand vibration, violent surges and overload, its action being independent of the nature of the gas. Dials having scales with divisions to 10 mbar, and capable of being rotated fractionally to offset changes in ambient atmospheric pressure, are available.

Diaphragm gauges, where a flexible plane shape of fine material separates atmospheric pressure from the vacuum connection, have

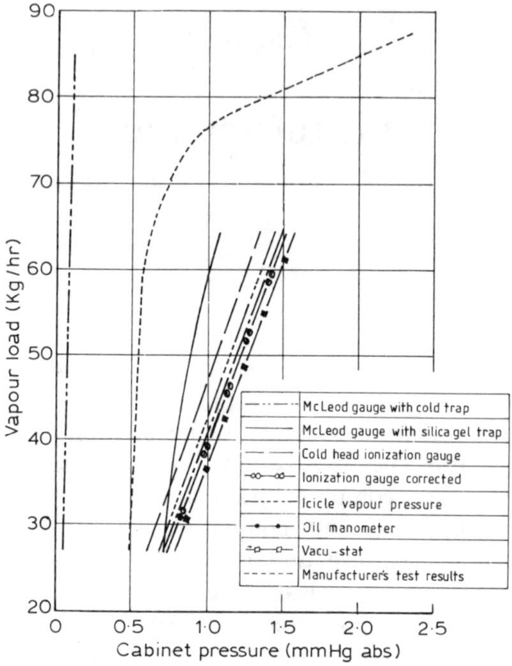

FIG. 49. Comparison of instrument readings during tests on steam-operated vacuum augmenters on AFD cabinet.

been commonly used to give a differential indication of sub-atmospheric pressure. Corrugated diaphragms, or stacks of such diaphragms, have developed into convoluted bellows, where much larger movements (passed on through linkages) are still linear with pressure. Sometimes a replaceable spring is incorporated which is then used to modify the unit movement per unit pressure and establish a particular *rate,* and therefore a sensitivity to a particular pressure range. A bellows capsule, permanently evacuated to a known (say 5 microns Hg) absolute pressure, can be linked to an identical capsule connected to the vacuum in the source to be measured, such that their differential movement indicates the absolute pressure of the source, *in a context* where 5/1000 of a mm of mercury can be considered as virtually zero pressure.

Alternatively, the vacuum connection can be admitted to the general interior of the gauge, where its pressure surrounds and acts on the hermetically sealed evacuated capsule. Within these two a further bellows system, at atmospheric pressure, permits linkage to the pointer. Thus a gauge covering 1020 mbar to '0' mbar can be companioned by one of similar dial size and scale length but reading 100 mbar to '0' mbar with ten times the precision

Extending this principle, a glass capsule sealed by a copper–beryllium diaphragm and exhausted to 10^{-3} mbar can react positively to the whole range from 1020 to 1 mbar, transmitting the flexure to a pointer on an individually calibrated non-linear scale. A reading accuracy of $\pm 1\%$ of scale length in the region below 50 mbar then gives 10 mbar to 1 mbar within $\pm 0{\cdot}25$ mbar, over a working temperature range for the instrument of 0–60°C. Such sensors are usually protected against vibration and sudden changes in pressure, by sintered filter, but response is almost instantaneous.

Instead of a purely mechanical linkage, a precision potentiometer can be used to convert the diaphragm movement into an electrical signal, from which paired calibration of the meter indicator is made. Such an arrangement can feed a chart record. A second potentiometer can identify (through the visual scale) and set a chosen pressure, actuating warning devices and/or triggering heavy-duty relays. Scale accuracy is as good at the expanded lower end.

From this design it has been a short step to remote digital indication, outputs for recorders, and switchable gauge heads operable at −20°C, based on flexible elements acting on a strain-gauge bridge.

Generally, however, beyond the level of 1 mbar absolute, the 'gas' needs energy fed to it in some way before its 'pressure', or ability to exert force on its surroundings, can be measured. Thus McLeod (or Kammerer) gauges, which trap a sample and compress it, are applying mechanical energy; others are designed to supply thermal energy, electrical energy or the energy of radioactive decay.

Electrical gauges can give continuous readings; moreover they can transmit their findings to recorders and, if required, by means of relays, can control automatically valves used in regulating the process.

In the especially significant range of low absolute pressure between 1 mm Hg (3/4 mbar) and 1 micron Hg, the number of gaseous molecules surrounding a filament of resistance wire carrying a constant electric current has a considerable effect on its temperature, owing to heat loss from conduction and convection. This functional relationship between the thermal conductivity and electrical conductivity can be an accurate measure of the pressure, even up to 10 mbar.

A Pirani-type vacuum gauge based on this principle contains a Wheatstone bridge circuit with tungsten filaments forming two of the arms. One is contained in a metal tube open to the vacuum to be measured. The other, identical, is in a compensating cell sealed off at a pressure below 1 micron. The bridge is balanced at very low pressure so that, when the sensing filament falls in temperature because the gas pressure there has increased, the out-of-balance potential provides a current which is translated into a meter indication of that pressure. Several sensing tubes can be used, with different scales.

In a variation, the filaments are kept at *constant* temperature and resistance by the heating potential, so that the bridge is always balanced, and the response so required is fed to the meter. Calibration correction curves for gases other than air are of course available.

The ionisation type gauge has a range from 10^{-2} mbar to 10^{-7} mbar and lower, and is thus mostly for use outside the sphere of foodstuff freeze-drying. It is based on the ability of electrons (emitted constantly from a cathode and directed past a grid which repels them) to bombard the molecules of residual gas in the evacuated sensor space, thus forming ions which then collect on the plate. The current so induced to the plate is proportional to the number of ions formed, which in turn is a measure of the absolute pressure in the system.

Simulator heads are available to check and reset, without actual vacuum, all the forementioned types of instrument, which of course are indirect, inference systems. There *is* one device that ignores the march of time, is direct and simple, and measures the absolute pressure of permanent gases easily within the range from 15 mbar to 10^{-3} mbar. McLeod compression gauges and their derivatives are for sampling the pressure in the vacuum chamber being investigated and do not permit of continuous measurement.

Commonly, by a swivel which allows the device to be moved from a charging position to a measuring position, a form of this gauge type cuts off a definite volume of rarefied gas at the unknown pressure of the vacuum system. The gas so trapped is then compressed by means of mercury in the glass gauge to a smaller volume at a higher pressure, the value of which is given by the difference in level between the mercury in that tube and in an adjacent compensating capillary. From a scale based on Boyle's law the pressure value in the vacuum system can be read off.

Sometimes a shortened U-tube is incorporated with the compression gauge to measure pressures as high as 150 mbar.

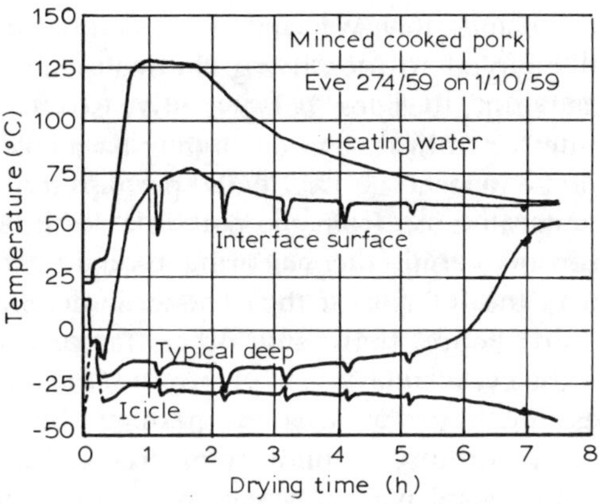

FIG. 50. Temperature graph of accelerated freeze-drying run, showing effect of opening up the sandwich of plates to weigh the product.

This family of gauges is very popular in industrial circumstances, although their accuracy, which is good, tends to fall off at very low values, and condensable vapours present an error problem, unless trapped out.

Temperature limitations are the guardians in freeze-drying (Figs 50 and 51). As an aid to action—whether it be the loading and unloading of an ice-condenser, the supply of cooling water, the evenness and speed of heating and/or cooling, the upper constraint with time tolerance of the dry material, or the residual ice temperature—some regular and instant check indication will be useful to monitor the steps.

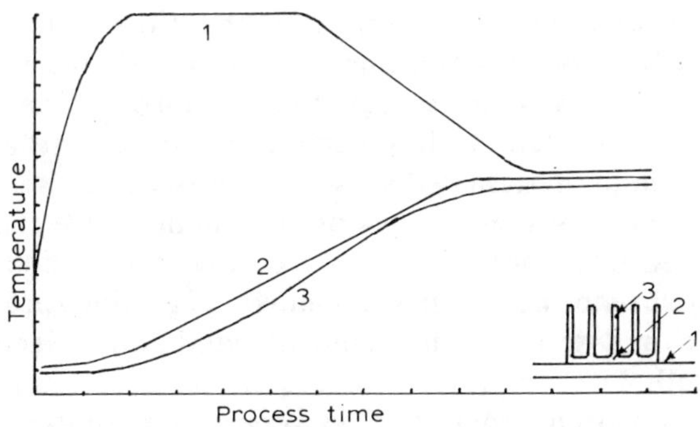

FIG. 51. Temperature profiles within aluminium ribbed dish during granule drying.

Instrumentation teaches that Wheatstone-bridge balancing, or slide-wire adjustment, coupled to identifying dotting pens, takes time to switch settle and print, besides being costly, so there has to be selection of the number of times as well as the number of positions and items of what may be measured. This leads (perhaps more so in a large batch plant) to generalisation from the particular of inferred information. Siting of sensors requires considerate thought. Information on what is going on in the interior of the product is dependent on these vital decisions and, more than any other factor, tends towards 'over-run' of process cycle time.

Thermocouples are popular sensors, perhaps because they are familiar already in laboratory or pilot plant pre-production investigations and trials. In themselves they can be remarkably neat; their compensating leads need not straggle, slendered against thermal side-effects yet not too fragile if properly respected.

These simple transducers in use with foodstuffs are preferably type K (NiCr–NiAl) and can be soldered to a metal surface for a point check, or to a foil or mesh 'spade' for local averaging, or inserted carefully in a token sample piece of the foodstuff. For this latter task, the physical positioning of the probe requires to be referenced, and its site as well as its tactual permanence assured during the entire process. A recommendable way of ensuring this being precise is to pass the junction down to the slant end of a (stainless steel) hypodermic needle, securing it in this sheath by epoxy resin or other inert filler. The sturdiness which is imparted enables the sensor probe, perhaps then no more than 0·5 mm in outside diameter, to penetrate non-frozen or chilled material precisely, and be securely frozen in. A 'gang' of such probes mounted on a handling block and arranged on an inclined plane has been used neatly over the cross-section of a meat slice particularly to assist in tracking the regress of the ice-front.

Contrariwise, it is virtually useless to stick a 4 mm sensor tip in a 9 or 12 mm thick morsel, as has been encountered. It is better to extrapolate from predetermined sensor information.

Tactile spade endings of stainless or aluminium gauze 50 mm square have been used in (wide) trays with typically in the order of 65 mm deep beds of shrimp, peas, chicken chunks, vegetable dice or similar, to gauge the mid-depth or ice-remnant condition of the randomly scrambled load. Their impressive picture of the temperature prevailing where they are frozenly locked in between sizeable pieces, which are assumed to be drying as a coherence, with sizeable interspaces, will be influenced by the proximity of the heating sources versus the vapour as it emerges. Experience may also dictate them to be buried horizontally nearer two-thirds down, but nevertheless they offer an end-point indication which has proved eminently serviceable.

Interpretation of signals from a thermocouple fastened to the surface of a heating shelf, say, on which a tray or dish may be deposited, likewise has to include consideration of thickness, spread of heat-flow and interface effects, before they may be declared representative indications of what transpires, and need regular inspection and care. Because they are 'always there' such attention tends to be overlooked.

The thermoelectric voltage generated gives an accuracy between ±0·3° and ±0·2° ordinarily, at an individual measuring rate to digital output of about $\frac{1}{2}$ s; automatic reference junction compensation for ambient temperature is standard, and commercial probe diameters are normally 1·5–3·0 mm.

Resistance thermometers, on the other hand, require an imposed constant current to give an equivalent voltage change in the temperature-sensitive resistor element, and tend to be thought of as for robust duty, in oil-filled pockets on heating circuits for example. They have been used (resin-coated) with spades in the product bed, and (surprisingly?) on refrigerated evaporator coils as tell-tales for the ice build-up. Permanently sited, with conduited leads to remote read-out stations, they are regularly incorporated in control sequences. Their less expensive (although comparatively heavy) connecting leads, freedom from intermediate and inadvertent thermoelectric junctions, and ease of integration with modern control circuitry, provided the non-linear temperature–resistance relationship is accurately known, tended to favour, no doubt unjustly, their industrial use in freeze-driers. Their role, however, is now gradually being taken over by thermistors, which have a negative temperature coefficient, i.e. the resistance increases as the temperature decreases. Also these are very much more sensitive.

Thermistor probes covering −60° to +30°C, for example, now available for immersion or point contact during freeze-drying, are $1\frac{1}{2}$ mm in diameter by 12 mm or 25 mm long, the first 4 mm containing the sensing area within a sheath of epoxy resin. Accumulated series resistance in the insulated extension flex of no more than 2 ohms will allow remoteness from control of some 60 m.

Mention has also to be made, in the context of the various sensing devices, of those for temperature which deduce pressure and/or pressure-sensing devices which deduce temperature. These are utilised in isolated or temporarily shut-off parts of a system, for instance to detect condensable vapours collecting in vacuum pump oil, and hence the need for purging or replacement. Again, pressure sensors of vapour production rates are used to deduce subliming ice temperature, as are resistivity or 'specific resistance' probes.

If the chamber is shut off from the means of vapour extraction, and the heat supply is simultaneously (as by raising dishes off shelves, for instance) curtailed, or otherwise prevented from interfering, and the air leakage rate is small and accurately known, then the vapour pressure in the chamber will equilibrate at that pressure corresponding to the instantaneous temperature of the ice front, which may thereby be inferred (Fig. 52).

Because of the several factors that seek to intrude on this somewhat delicate interpretation, the time for an observation, which will involve a number of readings, rapidly in succession, is limited and is prone to

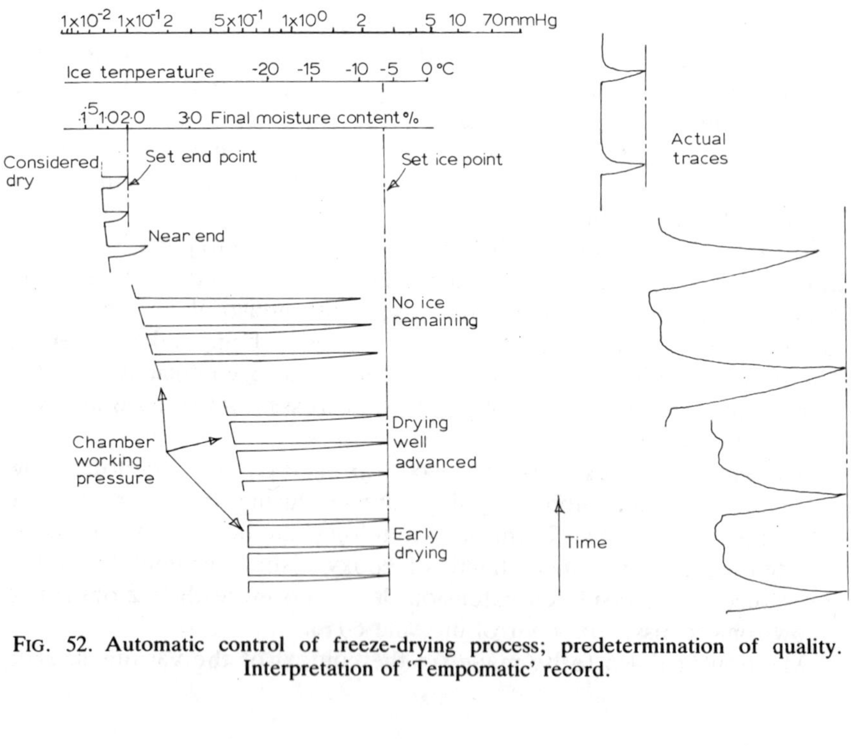

FIG. 52. Automatic control of freeze-drying process; predetermination of quality. Interpretation of 'Tempomatic' record.

more difficulty in a large cabinet than in a small unit. However, with computer logic at hand, these various factors can satisfactorily be taken into account. For instance, should the heat input fall away too rapidly, the inferred ice temperature at which sublimation *had* been operating will be rather lower than actual; should it remain long enough to intensify vapour production by warming sufficient ice, the balance will tilt the other way.

A succession of readings noting the pressure reaction over the whole isolation period will both indicate trends and warn against prolongation. Towards the end of drying more latitude is possible.

This 'hovering' method of measurement will result in a mean supplied from the entire load, compared with that from a limited number of (T/C) temperature sensors in preselected specimen parts of the produce, which requires generalisation from those particular spot-checks. The range of non-uniformity in most respects has perforce to be narrow.

A visual filing record of the process is commonly represented *minimally* as in Fig. 53, which endeavours to show (and usually *does* show) sufficient of the significant temperatures plotted against time in order

(a) to describe adequately what has occurred;
(b) to allow of planned repetition;
(c) to monitor departures from, and corrections towards, various 'set' points;
(d) to affirm the decision to terminate; and
(e) to enable a cam profile to be cut for regular control, *or* a semiconductor EPROM to be made.

Freezing may be done separately, on pilot or full-scale plant, in which case this part will be just as carefully followed, and the equally safeguarded transition episode allowed for.

While that surface of the product subject to prolonged exposure to (relatively high) dry temperature is seen to be the quality control criterion, the process may be so gentle and cautious that a representative heating shelf temperature is recorded instead. With hollow shelves, it may be the easier measurement of circulating fluid temperature, containing a broad assumption of by how much the various heat transfer coefficients insert lag and drop effects. The nearer to the 'true' surface condition one can dare approach, the more business opportunity there is of 'pushing' the process, even at a

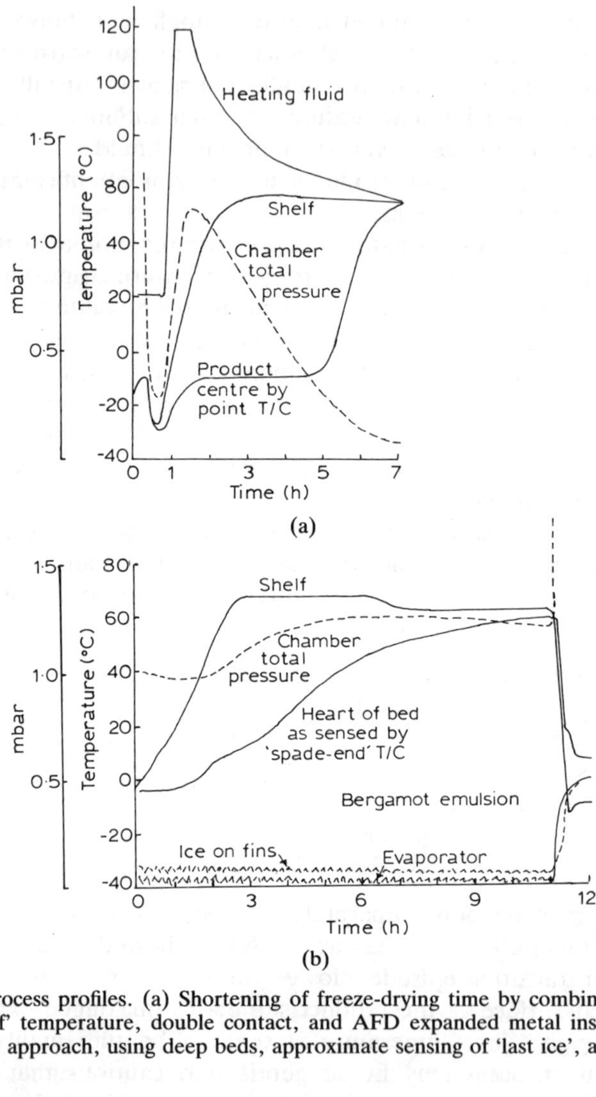

FIG. 53. Process profiles. (a) Shortening of freeze-drying time by combination of high initial 'shelf' temperature, double contact, and AFD expanded metal inserts. (b) The 'averaging' approach, using deep beds, approximate sensing of 'last ice', and over-run.

modest level. Caution based on lack of (or imprecise) information is tacitly inefficient.

Not necessarily to a comparative scale, the (sub)-atmospheric condition within the drying chamber may be recorded as the (total) absolute pressure, or the partial pressure of non-condensables, or tidily as a signal which simply indicates, without the actual value being regularly recorded, that the level is sufficiently within sensible limits for other features to be read or indeed to take place. In this spread of alternatives the parameter of 'vacuum' is unique.

Ice-condenser temperature, like appropriate product temperature, represents a limit or constraint on the process. It may be read as evaporator refrigerant temperature, or tube external surface spot temperature, extremity of fin or even some proximate position to which the accumulation of ice or snow will aspire—signalling a need for defrosting or end-of-process adjustment. In the nature of things, as the rate of production of vapour falls off, so will the demand on the refrigeration machine, which will compensate by lowering the evaporation temperature and/or shutting itself off. The record (Fig. 53(b)) will exhibit an oscillating or even a 'bumping on the bottom register' effect, depending on how it is controlled.

So far the siting of these parameter measurements is possibly, for the most part, in the consummate gift of the equipment supplier, but the *sensing* by placement of what is to happen and what has actually happened, within the particular product, is very much (and necessarily) in the hands of the operator.

Two pieces of information which the foodstuff in depth must yield are (a) whether there is ice remaining, and (b) what the highest temperature of that ice is. This is not quite the same as asking if the foodstuff is completely frozen, because for the complexes of which it is made up there will be several waters segregated as ice from a number of constituents, and freeable possible waters still trapped or held with most of these in stronger solution. Depending on its history, this further 'water' may be beyond the intention of the procedure or the capability of the extraction process to remove. On the other hand, there undoubtedly will be OH groups which can by persistent extenuation be extracted irreversibly and should not form an acceptable part of freeze-drying, although such circumstances may unwittingly happen. Some proteins could have been deactivated in a preparation procedure, being denatured in such a way that renaturing can occur because of their ability thermodynamically to conformate. Some proteins may be denatured by the thermal exposure caused by the effort to introduce the heat of sublimation through the already dry tissue, and prolongation of heat introduction after the last vestige of readily available ice has been removed merely increases such risk at the parts which have been longest dry.

As the product dries towards the remotest interior where a thermocouple may be sensing, its hold on the stem relaxes, and care must be taken to support any external weight of leads so that it or its neighbour is not dislodged. Again, should an artificial icicle containing a thermocouple be used to determine the partial pressure of water vapour in the general space of the chamber, with which its sublimation rate and therefore temperature will be in equilibrium, this arrangement is finite; it is exposed to general radiant heat and may be lost by loosening during a run. It is provident to check over such items at close-down routinely, even if no error is apparent.

Evenness of heating is considered under methods of heating.

INFERENTIAL DEVICES TO DETERMINE THE END OF DRYING PROGRESS

(a) Computer Use of the Sorption Isotherm

From (e.g.) Henderson's equation (1952), one obtains

$$X = \sqrt[B_1]{\frac{\log_e(1 - a_w)}{-B_2}}$$

where X is the instantaneous average moisture content (% dry basis), a_w is the water activity (i.e. the instantaneous Equilibrium RH/100) and B_1 and B_2 are statistical parameters describing the predetermined isotherm. Thus, assuming the data can be obtained and relied on to satisfy the equation under the relevant temperature conditions, the tentative method works back ingeniously from a measured humidity in the freeze-drying 'chamber' to deliver a succession of numbers for X which will gradually tail off to a steady value. The reality of the actual figures matters less than the evidence of trend.

(b) Isolation of Drying Chamber

Periodically, a valve between the drier and the chamber holding the evaporator coils is shut and the absolute pressure in the product area is read. Any rise in pressure due to leakage is allowed for from previous data. The rise in pressure due to vapour alone, heading towards equilibrium with the temperature at the ice-front, will indicate the rate at which sublimation is instantaneously occurring, and therefore the existence of residual ice, for instance. This procedure must not last too long, or the heat in the dried portion will alter the temperature of the ice. Again, the initial rate-of-rise comparisons are more valuable than actual interpretation of the numbers.

(c) Icicle on Thermocouple Suspended in Chamber

Here pure (distilled water) ice, protected as far as possible from heat radiation, will sublime to be in equilibrium with the vapour pressure in the chamber; as it recedes, so will the equivalent temperature as measured (and reported) by the thermocouple.

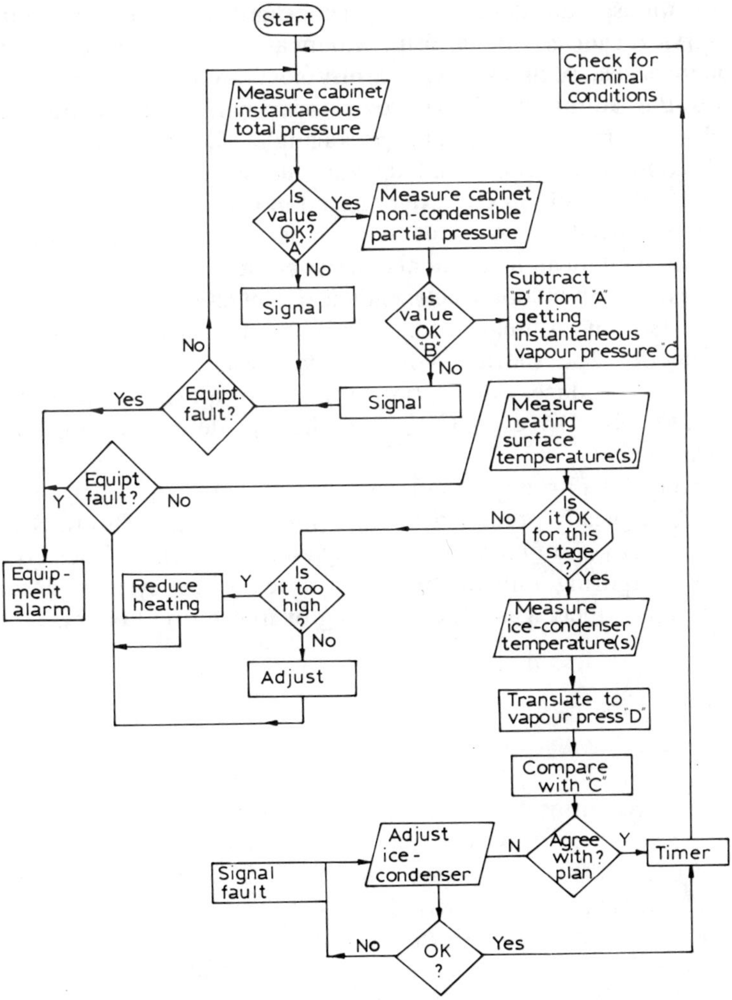

FIG. 54. Flow-chart.

Whatever form of sensor or range of sensors is chosen for the task of supplying the various pieces of intelligence demanded in order to take full control of the process schedule, they will all have to be capable of incorporation in a flow-chart system. The sequence of operations to be carried out by a computer monitoring a dictated process will be in a series or cascade of loops as illustrated heuristically but simplified in Fig. 54.

The widespread desire among supermarkets to exhibit, on chill self-service shelves, fresh fish and meat portions enclosed in see-through packs, perhaps having flimsy covers on rigid tray bases, has led to the adoption of 'controlled atmosphere packaging' (CAP); MAP = 'modified atmosphere packaging'. The head-space of such retail containers is flush-filled with air components in abnormal ratios—high usually in CO_2, diminished in O_2.

The beneficial technical connection is that CAP has prompted a series of analytical instruments designed for rapid response to gas mixtures, based on an electrochemical 'detector' of reasonable life. The modern micro-fuel-cell sensor can monitor O_2 levels from trace to atmospheric. The thermal conductivity method, complete in about 1 min, can produce (with print-out) the proportional quantities of CO_2, CO, N_2, O_2, even CH_4, and thereby detect leaking packs or bacterial activity.

Commercial nitrogen is not entirely oxygen-free, and freeze-dried packs are typically preferred to have not more than 2% O_2. Sampling is portable, by the fine needle of a hypodermic syringe inserted in the septum, requiring only 5–10 ml for a test. The system is suitable therefore for 'inert' gas head sampling of freeze-dried packs.

Chapter 10

Packaging

Packages for freeze-dried foodstuffs properly must provide, above all other requirements, timely and timed protection from ingress of moisture and oxygen. In fact, the package as the consumer finds it should provide a portable, safe environment in itself, contrasting with the need for machinery back-up to the frozen food chain. Temperature consideration for storage of dried foods is fairly flexible.

Generally the protective container should be of handleable size, perform well on filling machines, be easily identifiable and convenient not only for smooth service in the factory, in storage and distribution but in end use. Visual aids to marketing may be restricted to an outer case. This overwrap may form part of the climatic barrier system, but additionally and *not* usually to the extent that the individual unit within is markedly scrimped. Corrugated fibreboard cases of a size suitable to be palletised are the most common protection against crushing.

Packaging costs, a quality problem, are not always held to the minimum necessary simply to convey an item, since they represent the means essentially of how it is to be sold. Needless to say, well established containers are utilised. Cylindrical open-top tin-plate containers and seamless-body aluminium cans/trays/dishes are universally popular. The inert cleanliness of glass, especially in jar form with heat-sealed waxed paper or foil cover, and follow-up replaceable lids, is attractive for products dispensed at random intervals over a relatively long period. Opacity is not often stipulated as an intervention, although wrap-around labels may cover most of the jar.

Flexible pouches tend to remix components which separate out during storage, such as dry soup ingredients of different densities and

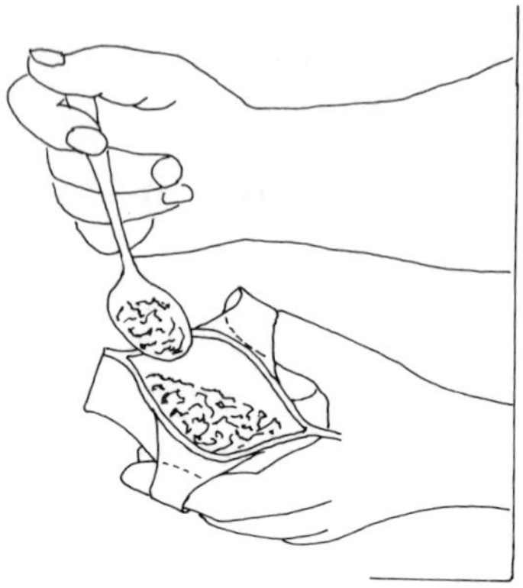

FIG. 55. In zero gravity conditions, the foodstuff adheres to both sides of the spoon. Picture indicates scraping the back of a spoonful on the lip of the pouch, held open by means of loops.

dimensions. The reconstituting liquid is sometimes measured directly into such a pouch, with stirring or kneading occasioned by flexing the bag (Fig. 55). Multi-wall sacks are used as intermediate storage.

Depending on the physical nature of the product, either volumetric (piece-counted) filling or bulk/fine feed cut-off 'weighing before filling' may be used. Give-away is electronically controlled. Head-space content is stipulated carefully by inert-gas flushing, vacuumising, packing room atmosphere, but rarely by catalyst or absorbent. Sometimes a homogenising (extra-dry scavenger) ingredient is employed.

Composite barriers of laminated plastics are much more popular in a variety of packaging tasks, other than for freeze-dried goods, and have therefore become much cheaper. The general requirement comprises an inner material which is heat-sealable, an outer which is colour-printable, plus—where neither of these substances is sufficiently adequate—an intermediate functional barrier (say of aluminium foil) which is gas-transmission proof as well.

Hazards include (1) seam/closure failure, (2) porous (paper) or pinholed (foil) materials, (3) diffusion through some plastics, and (4) contamination from adhesion means (now virtually eliminated).

Permeability of a flexible packaging material, as some sort of lightweight film say, to the twin scourges oxygen and water vapour is commonly quoted, in the form of test results, by the manufacturer.

Thus, respectively, the units may be (a) cc O_2/m^2 per 0·025 mm thickness per day at 1 atm partial pressure difference and 75% RH, at 20°C, and (b) g H_2O/m^2 per day similarly.

Such tests are on uncreased samples, not made up. Caution is advised in extrapolating 'per 24 h' beyond the period of the test. Likewise thickness for linearity of permeability should be between limits. Of course a partial pressure difference of 1 atm, while a standard base strictly for comparison, and short list selection, seems unreal, since (even if the dried foodstuff accepts the oxygen completely as it arrives) a constant difference of any magnitude is unlikely in practice. Nor for that matter is 100% RH a typical condition, but rather a specific requirement.

Cost tempered with precautionary common sense will dictate the brinkmanship in barrier decisions, while some form of accelerated storage test, in full 'mock-up' will penultimately help to confirm the selection.

Specimen Data

Oxygen tolerance in the N_2 head-space for sample dry product, found organoleptically just acceptable after 10 weeks test storage at 25°C, (say) 0·1 mg O_2/g dry (cooked beef, minced).

Single portion of 65 g occupies 100 cc in a pouch of total (back and front) surface area 375 cm^2.

Proposed material for pouch: 12 μm polyester metallised on one side, adhesive laminated to 75 μm low-density polyethylene, such that the metallised layer is encapsulated.

As a barrier to oxygen this permits (say) 1·0 cc/m^2/24 h at 25°C to pass when there is a pressure difference of 1 atm of O_2 across it, and therefore only 0·23 cc/m^2/24 h at 25°C when only the O_2 content of normal air is encountered.

Then, assuming density of O_2 at NTP is 1·429 g/litre = 1·383 mg/cc at 25°C, the time to pass 6·5 mg O_2 through 0·0375 m^2 of material (which in 24 h lets in 0·0086 cc of O_2 at constant pressure difference and constant 25°C) will be

$$\frac{6{\cdot}5}{1{\cdot}383 \times 0{\cdot}0086} = 546 \text{ days or nearly } 1\tfrac{1}{2} \text{ years}$$

which is not only longer than the 10 weeks to taint at full exposure, but also of the expected life of the pouch from factory through point of sale to camper's actual consumption of product!

The above rough calculation is included as evidence of blithe assumptions fraught with dubiety, which must not be left there but tested out scrupulously. Its main advantage is that it tends to transfer attention to the MVTR properties.

The converters of the above laminate give the 'moisture vapour transfer rate' as $<0{\cdot}8$ g/m^2/24 h at 38°C, 90% RH (an air condition of 0·039 g water vapour per g dry air; quite severe). Through 375 cm^2 this is 0·03 g/24 h.

Assuming that the foodstuff is at 1% moisture content on a dry basis, the amount of bone-dry material is $(100/101) \times 65$ g $= 64{\cdot}36$ g.

Taking the safe upper limit (say) as 7% moisture content (dry basis) $= 0{\cdot}07 \times 64{\cdot}36 = 4{\cdot}505$ g, the amount of moisture to be added to just reach this is 3·865 g, and the theoretical time for this to happen is $3{\cdot}865/0{\cdot}03 = 128{\cdot}8$ days or 4 months. The two 'enemies' will of course work in conjunction.

Packaging materials for freeze-dried foodstuffs are indeed multiform. Their scope stretches from elaborate composites capable of surviving several years diurnal variation in a tropical rain-forest to primitive coverings calculated just to last a local distribution.

Partly the trend stemming from the former extreme outlook has been a limiting but immediate delay, because of a stream of objections and queries produced by competitive lobbies. These largely concerned proof of compatibility with foodstuffs of various laminate adhesives.

The ensuing prolonged series of tests under the auspices of the FDA gradually released on to an avaricious expansion of the market a proliferation of alternatives. Many of these were of interest to processors of other than freeze-dried commodities, so the costs due to volume came comfortably within general reach.

As for the latter extreme, interest there also developed to a volume market, mainly because the attributes of freeze-dried foodstuffs which captured the imagination lay not with longevity but with rapid reconstitution; retention of flavour came a close second.

Contract freeze-drying of customers' prepared substances, supplied and returned in bulk, came into vogue, such that the finished freeze-dried goods made their journey from the 'primary' processor to the ingredient assembly processor in multi-wall sacks or the like. If the ingredient became part of a yoghurt pack there and then, for instance, it was reconstituted as a recognisable item in the *liquid* pack, to which a different type of packaging specification pertained. Otherwise, if still

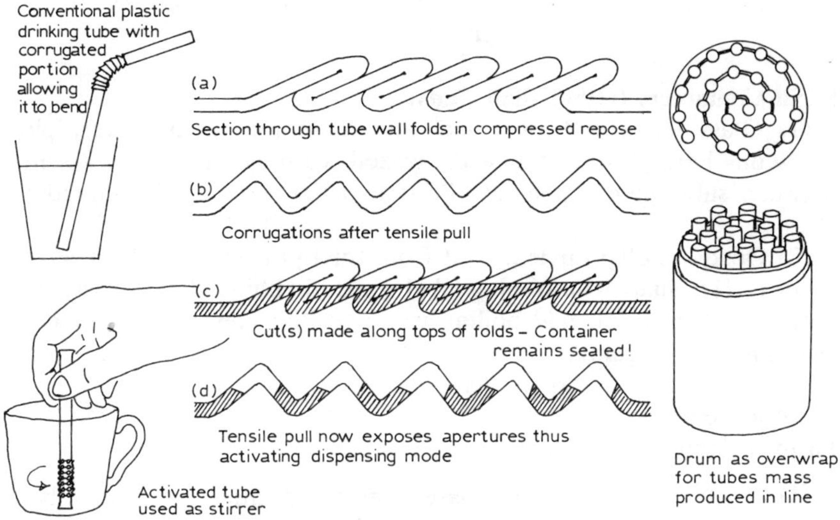

FIG. 56. Custom-designed container/dispenser/stirrer for soluble solids of freeze-dried tea extract; individual portioning; colour-coded for other contents.

dry, it would be made up into portable environment 'permanent' packs whose durable nature and tolerance were arranged to suit the intended use (Fig. 56).

Thus 'boil in the bag' packs having actual holes have been used as inner containers within a tight 'whole meal' main pack. Picnic pouches into which a measured quantity of hot or boiling water could be poured, the dry ingredients being stirred into a soup (say) *in situ,* are another form of the same idea.

Some breakfast 'cereal and dried fruit' miscellanies use freeze-dried sizeable pieces of fruits for intense colour and full flavour aspects, along with ordinary dried vine fruits, in what are virtually waxed paper inners, card cartoned. In such cases, the proportions of dry cereal to fruits are in themselves a protective 'moisture-seeking' surround, and the contents of even an opened then mouth-folded bag retain their freshness for an adequate period.

All this is certainly not to say that packaging should be approached in carefree fashion, merely that the industry now has experience of a complete range of protective wraps from which judiciously to select.

PACKAGING FORMS (ALREADY WELL KNOWN TO THE PUBLIC)

Rigid Containers (Other than Cans)

(1) Glass jar with heat-sealed aluminium foil mouth-cover, plus screw-off hard-plastics lid with waxed card joint for protection between subsequent openings. Printed paper wrap-label adhered to glass.

(2) Opaque polypropylene (eat from this) pot, with similar peel-off cover to (1). Shape-conforming label trapped behind clear polystyrene outer. 'Pull back lid, add boiling water to "fill" level, stir, re-cover, wait 4 min, stir again.'

Flexible Pouches

Form/fill/seal machine range:

(1) E.g. 'eat direct' type, based on composition of 'boil-in-bag' pouches—a laminate of polyethylene with oriented polypropylene coated with vinylidene chloride copolymer.
(2) E.g. potato crisp (chip) type, where no heed is paid to a life expectancy of more than 6 weeks, and the safeguard is therefore an outer overall more substantial sealed protective cover. Usually oriented towards individual portion group distribution.

Micro-fuel-cells are now available, as a spin-off from Space research, to measure the concentration of oxygen in a sample gas, and therefore have potential for package testing. This sealed electrochemical transducer is specific to oxygen, with an output linear with oxygen concentration from low ppm to 21% (see also p. 176)

Normal wholesale and retail grocery storage conditions for dry goods may be taken on average in the UK as 15½°C and 60% RH (0·0065 g moisture per g of dry air). Packaging permeability comparison tests may be carried out in a humidity chamber set slightly more severely at 25°C and 75% RH (0·015 g/g).

Although it is provident and proper to use complete barrier laminates, some products are packed in limited permeability wrapping material which can hold back deleterious moisture (vapour) at estimated ingress rates for longer than the planned life of the pack. Such apparent brinkmanship is carefully costed in terms of calculated risk.

The range for selection of plastic laminate pouches, identical in size and shape, are prepared with a known weight of known moisture content (M_0) foodstuff sealed inside. A similar sample of the foodstuff is allowed to equilibrate with the conditions in the chamber so that M_H at H may be calculated.

Then, from $dM/dt = c(H - h) = k(M_H - M)$ nearly, after time t (days), it can be shown that the half-value period for a package will be

$$T = \frac{\log_e 2}{k} = \frac{\log_{10} 2}{\tan \phi}$$

where a few plots of $\log_{10}(M_H - M)$ against t give a straight-line graph whose angle to the t-axis is ϕ (Fig. 57). Comparisons of T found for each type of package are then overall measurements of each one's efficiency, independent of the moisture content of what is inside.

In-container desiccants have been used to finish drying out to very low moistures (Hendel *et al.*, 1958), and the concept of pressure gradient steps in a quasi-steady-state operation governed by moisture transfer coefficients has been utilised to estimate, roughly but ade-

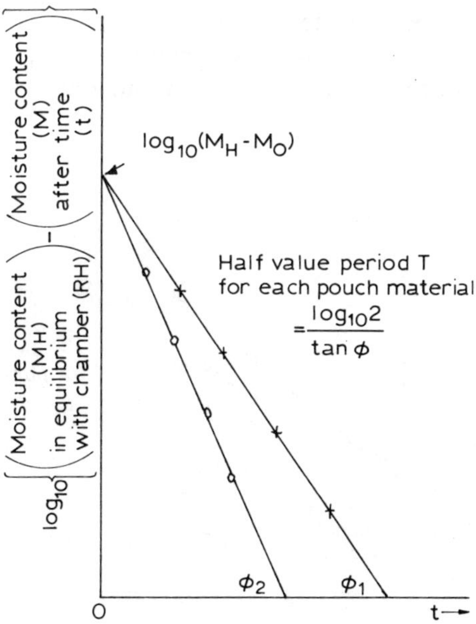

FIG. 57. Pouch material permeability test; efficiency comparison.

quately, the time to equilibrate. Thus

$$\frac{Q}{T}=\frac{p_\mathrm{p}-p_\mathrm{d}}{\dfrac{1}{k_\mathrm{p}W_\mathrm{p}}+\dfrac{L_\mathrm{a}}{k_\mathrm{a}A}+\dfrac{1}{k_\mathrm{d}W_\mathrm{d}}}=\text{overall rate of transfer}$$

where p_p is the average water vapour pressure of the ingredient with moisture to give (mm Hg), p_d is the average water vapour pressure of the ingredient scavenger of moisture (mm Hg), k_p is the average transfer coefficient (g/day . cm^2 . mm Hg), W_p is the weight of donor (g), L_a is the average transfer distance through the package atmosphere (cm), k_a is the average transfer coefficient through this atmosphere (g/day . cm^2 . mm Hg/cm), A is the cross-section of the container (cm^2), k_d is the average transfer coefficient to recipient (g/day . cm^2 . mm Hg), W_d is the weight of recipient (g).

Such vapour transfer coefficients have been found to vary within limits, falling with moisture content and distance.

More often several ingredients, all acceptably dry but of various degrees of dryness, are mixed to compose the pouch contents, and then it is not so much external moisture as migration within the package that has to be considered (Fig. 58).

Even with only one substance, weighed out from material batches of differing moisture content, some homogenising has to occur.

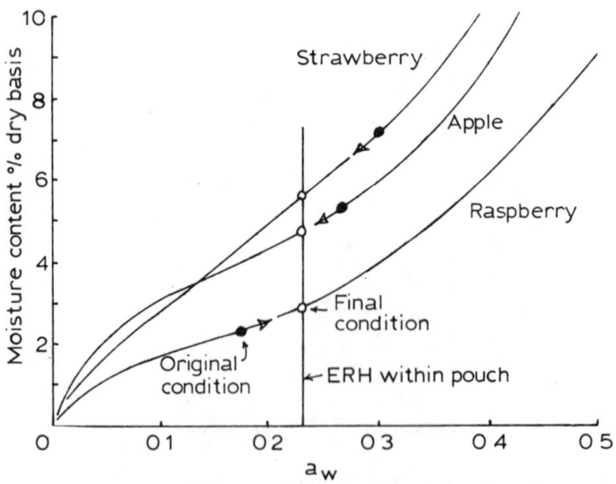

FIG. 58. Use of isotherms to determine harmony within the pouch.

Averaging will then simply produce

$$M_{\text{total}} = \frac{M_1 W_1 + M_2 W_2}{W_{\text{total}}}$$

The latitude between first 'dried' and last 'dried' portions or particles in a tray or batch may be relatively small, but will adjust over time in this fashion.

Commercial pressures may dictate that, while the last vestige of ice must disappear before the run is officially terminated, the surfaces long-since 'over'-dried will take care of migration within the piece, to arrive at an overall planned evenness of moisture content.

Sometimes a scavenger ingredient, i.e. a foodstuff that has been deliberately dried down to a moisture content well below the eventual homogenised final value (such as may be arranged by allowing the end-of-drying temperature to rise throughout the mass), is included as a means of 'finishing off' the drying of other ingredients during the longer (and cheaper) storage period.

IMPROVISED GLOVE BOX

A useful inexpensive adjunct for the packaging examination laboratory, when scrutiny of a_w in the head-space of soup-sachets, cup-stack sleeves, coffee jars and the like is called for, is an ordinary disposable polyethylene glove (Fig. 59), as is commonly available for household or medical use. Although the polythene may be only 0·025 mm thick, and the idea therefore something of a heresy, its permeability rate to water vapour *in the context* is of no consequence for a test lasting a mere 15–30 s. Individual cases require discretion. The fingers are utilised to hold implements—knife, humidity probe, sealing tape—sealed off from laboratory air by tight binding, and brought into play as needed. The package to be examined is inserted and sealed at or into the wrist-hole, the glove being squeezed empty of air as much as possible. The clear flexible plastic enables the package to be cut precisely, the probe inserted, readings of temperature and humidity taken which are steady, and the cut subsequently sealed, all with minimum intrusion and margin of error.

Even when put over the foil-covered mouth of a glass jar containing say 200 g of freeze-dried coffee granules, where the ERH at 20°C is <5% ($a_w = 0{\cdot}05$), this is a practical *ex tempore* device. At that level,

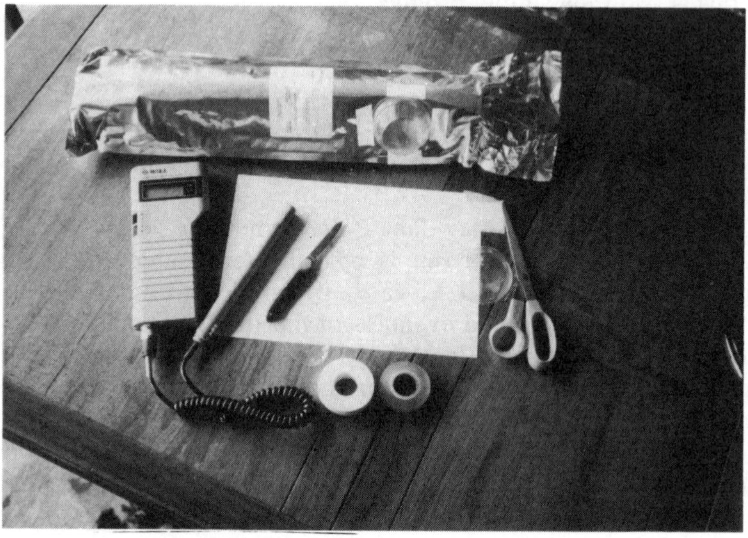

(a)

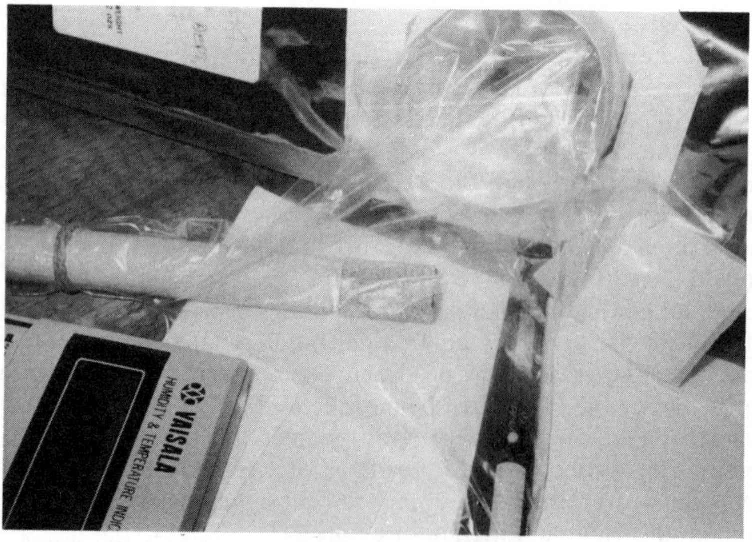

(b)

FIG. 59. Improvised glove-box. (a) Humidity and temperature probe, with knife and sealing tape, all to be accommodated securely in the fingers of a plastic glove to be sealed onto the package under scrutiny. (b) Close-up in position; speed minimises error.

the EMC is about 1·0% from the isotherm, which means that there is a total of <2 g moisture associated with and throughout the actual coffee, and some 0·001 g total in the interstices and head-space. The probe reacts promptly to an equilibrium condition in the head-space and the surface layers of the mass of granules.

Incidentally, if the probe is wormed into the bulk of the dried coffee, no change of reading is discernible, even if the jar mouth is then exposed freely to the ordinary air. This is in line with the common usage experience of consumers during a week or two of frequent opening and closing of the jar. Although the very surface layer of coffee granules may react promptly to the exposure, the main body of granules does not seem to 'suffer' measurably, as it soon would within a day if laid out in a single layer. So the diffusion intensity over the relative time spreads the hazard, away from the apparently more vulnerable surface layers which in consequence do not 'cake' in the bulked context. This can only be because the interstices are such as to allow freedom of instant and even dispersal.

The deflated glove adds no significant room air behind the inserted probe, either on initial piercing or during the time to settle the reading.

The sigmoid character of most curves shows that the moisture content towards very low humidities remains greater (takes longer for a_w to fall) for foods like egg-white and potato, whose dry solids are richest in protein, starch or other polymer of high molecular weight, than it does for foods high in soluble solids like apple or cabbage, relatively, emphasising the difference in water-holding location and gelatinous relationship. Where there is a 'straight middle' to the curve, behaviour is easier to calculate.

With mixtures of dried ingredients prepared separately, and having different EMC/ERH characteristics, the simplest way of arriving at the effect of combination is a pragmatic sampling technique as outlined above, which gives a reading on the high side. Reference to the new position on respective isotherms will, knowing the amount of each component entered to the mix, enable the various moisture transfers (of zero total sum) to be worked out. None should exceed its particular 'safe' limit.

One specially over-dried component may be used deliberately as a scavenger. Forecasting what is likely to happen is dependent on an assumption of constant slope of each isotherm over the relevant portion, thus

$$\text{Resultant } a_w \text{ of mixture} = \frac{\sum a_w \,.\, S \,.\, m}{\sum S \,.\, m} \tag{1}$$

where a_w is the initial a_w, S is the slope of the component within range, m is the mass of dry solids in the component.

Again, over a small moisture range, it is possible to assume a quasi-steady-state rate of desiccation between the adjacent mass of one component towards a solitary other, which may be treated as the product of an overall mass transfer coefficient [in units of (say) g H_2O/day . cm^2 . mm Hg] and the average water vapour pressure difference (mm Hg) times the cross-sectional area (cm^2) of the interface. Knowing the temperature, p_0 can be obtained from tables, and hence each 'p' from its ERH.

Including the package atmosphere changes the units of the related coefficient to g H_2O/day . cm^2 . mm Hg/cm, since the difference in the water vapour pressures at each interface has to be divided by the average transfer distance 'x' through the package atmosphere, including presumably half or so of each product depth. For each product amount, a similar relationship may be defined, the coefficient then being in some such units as g H_2O/day . (g product) . mm Hg. Thus

$$\frac{dM}{dt} = \frac{p_A - p_B}{\dfrac{1}{k_A W_A} + \dfrac{x}{k_a A} + \dfrac{1}{k_B W_B}} \tag{2}$$

from which a time required (taking incremental steps) to achieve a given moisture transfer may be estimated, provided of course that the various transfer coefficients can be determined.

These appear to range, by a factor of at least 10, over typical equilibrium vapour pressures, so that the standard transportation method of treatment, where the magnitude of the flow is always directly proportional to the potential, as well as to the magnitude of the appropriate constant of proportionality, is rather awkward to manipulate. Nevertheless, it is interesting as a basic exercise in circumstance planning.

On similar recognisable lines, when a substance is exposed to a (hostile) imbalance with an atmosphere, either directly or indirectly through a permeable container, then

$$\frac{M - M_e}{M_0 - M_e} = e^{-Kt} \tag{3}$$

where K specifies the transfer rate, M_0 is the m.c. at start of test ($t = 0$), M is the m.c. at time t, M_e is the m.c. in equilibrium with the atmosphere.

Neatly to tackle this, suppose M is the mid-point condition $= (1/2)(M_0 + M_e)$; then $1/2 = e^{-KT}$, so that the half-value period is

$$T = \frac{0{\cdot}692}{K} \tag{4}$$

From eqn (3) one obtains

$$\log_{10}(M - M_e) = -\frac{\log_{10}2}{T} \, . \, t + \text{const.} \tag{5}$$

Hence a plot of $\log_{10}(M - M_e)$ against time t will give a straight line of slope $-\log_{10}2/T$, from which T and then K may be found. Note that the method adapts to an increase or decrease in moisture content (m.c.), also that, provided the moisture sensitivity of the wrapping material is taken into account by suitable conversion of its weight to a proportion of foodstuff wrapped, then direct weighing can be used without determining the actual moisture content of the foodstuff itself. For quality control routine procedures, a number of packages may collectively be weighed at suitable intervals. Individual laboratories arrive at the most convenient systems of dealing with their habitual range of goods by amassing carefully a library of isotherms which are sufficiently averaged in accuracy to be perfectly adequate for their responsible needs.

Part IV

FUTURES

Chapter 11

Reconstitution

RECONSTITUTION—THEORY AND PRACTICE

There is an assumption that freeze-drying a foodstuff or beverage implies a distinct later stage of restoring the moisture removed before eating or drinking. This is not necessarily so; astronauts have been trained to find salivation adequate preparation of a number of substances for satisfactory and enjoyable swallowing. But more commonly it is expected that the substance be actually seen to be restored to something closely approaching and resembling its original moisture-laden condition if only to reassure the partaker that the process is fully reversible and the (long) period in storage limbo has been wiped out from its history—as good as new.

This rehydration, adding water to a dry (i.e. dried) substance in order to reconstitute it, is strictly not the reverse of the process which removed the moisture, although in many instances it suffices for the end in view. When the material was brought to the frozen state, the H_2O to be removed was separated from the complicated chemical constituents within and around the cellular structure (*sic passim* siccative, as it were!). It (and in most practical circumstances, only it) was then removed by change of state to a much freer arrangement of the (moisture) molecules which in turn enabled it to find its way to the surface and escape. This overall escape was from a fabric of locked constituents which themselves retained for the most part an unaltered structure.

The intricate labyrinth so derived has been simplified and stylised almost out of recognition for the sake of easier mathematical analysis, until the foodstuff is at fault for not conforming, and the theory can proceed unfettered. Fortunately the labyrinth does not remain forlorn and empty but is filled with other molecules in a gaseous state. Clearly, complete reversal of this procedure is inherently difficult, if not impossible, and certainly not to be viewed as convenient.

Blithely the eager consumer simply adds water (any water), hoping for the best, and amazingly enough this simplicity nearly always pays off, producing in a surprising fraction of time, even in the case of sizeable pieces, a close resemblance to the original in colour, texture, odour and flavour (Fig. 60).

As a means of hastening the reconstitution, the water may be warmed, and the larger pieces floated on it, on the premise that the air or whatever in the myriad of interstices will yield place to the water surface advancing because of capillary effects, or just the gravitational weight of the piece bearing down on it. For material which has been vacuum-packed (and therefore has less gas in the spaces) it has been known to dip the corner of a pouch under water before snipping it off, thus inviting the water more readily to enter the foodstuff.

FIG. 60. Melting snow over a small spirit stove in preparation for reconstituting dehydrated rations. 'Public Information—United Kingdom Land Forces'.

All these procedures work, with less apparent bother than the description might suggest. Frequently freeze-dried materials are mixed with other ingredients, such as modified starches, for quick soups and the like, and their wetting characteristics overshadow those of the freeze-dried items, whose major role may be to catch the eye with colour and sizeability. Inert-gas filling to atmospheric pressure has largely overtaken vacuum packs, and in any case the outside of such a pack would presumably have to be assured clean before being contacted with the water of reconstitution which is destined to become part of the foodstuff.

It is this water itself which tends to be overlooked or passively ignored as an active unassuming agent. Water supply authorities commit themselves to little more promise than that their commodity is 'potable', i.e. 'fit to drink', and therefore contains less than some upper limit(s) of a variety of noxious substances.† In fact, in the firm belief that water should have a 'taste', in the knowledge that distilled pure H_2O has no taste, and faced with the immense unpopularity which the expense of further elaboration would call down upon them, the purveyors of this universal communal prerequisite actually advertise and extol the differences in it to be found in various geographical regions; in truth, people grow accustomed to their local water supply, especially in remote and deprived areas and circumstances, and even prefer it to that from elsewhere. So they rarely object more than tokenly to a reconstituted dry product on the grounds that the water with which it was mixed was not in any sense the exact replenishment of that which had been extracted. Their chemical awareness may not be quite so deep as to appreciate that a colloid unable to leave a dehydrated cell may not be glad of an H_2O burdened with salts endeavouring to 'rejoin' it. Albeit this is the manner in which such intrusion is promoted.

Physically, this is probably one of the main mechanisms that operate during reconstitution of a substance whose cell walls retain resilience and still act as semi-permeable membranes. The accompanying salts

† That ancient god of measure, the hydrogen ion, which smiles obscuringly on balanced dosage of acid- or alkali-producing additives, as offerings of potability if not palatability, may also be appeased by judicious use of natural resins. Inorganic cation-exchange adsorbents (such as fullers earth) may be followed up by a regenerative common salt (sodium cycle) treatment which demineralises without removing some anions. Thoughts along these lines may lead to the sacred cow(!) by-product(s) mentioned on p. 212.

remain outside the cells, as radicals let us say, while there is a true swelling of the cells, refilled without regurgitating, up to 98% of the liquid commonly being retained.

Where natural waters are bottled and sold, they tend to be marketed as heavily charged with minerals and, if recommended for their therapeutic value, are consumed by themselves as a dose regardless of taste; not for them the preliminary conjunction with food outside the body. Likewise, although distilled water may be used to reactivate skin-graft material or blood plasma yet not milk plasma, after freeze-drying, time is not of the essence, and existing body fluids are called into play not long afterwards. But rapidity is a keynote as well as simplicity in preparing convenience foods, so tap water is normal, and fortunately so.

Halogen presence where it occurs would be too faint to be used as a tracer in establishing the role of water-of-reconstitution constituents. Not much has been done to determine exactly what interchange (other than by weight) takes place; the evidence usually has no more than transient interest. Rarely, if at all, is there mention of the hardness or other property of the water.

In England, a land of bountiful rain, where village sites at one time may have been decided by water-supply, emerging below the chalk and above the clay, industrialisation has grouped large centres of society around the backbone, respectably remote from the wet western and other uplands which mostly form their catchment areas. They in turn treat their waste water to free it or neutralise it from passage through the system to an ordained rather than a preordained level of acceptability. A similar hiatus occurs in the precincts of all ordered communities, and several sequences may occur to the water in its journey to the sea and the clouds again. Beyond that 'sink/summit' cycle of expediency there have been attempts to reclaim solutes under the aegis of international bodies, although generally the next community to draw water from the gravity line has little recourse but to accept what comes to it. Thus, unless very elaborate means are taken to go direct to the (water) table, for a lot of people their 'fresh' water has already been recycled in intimate fashion!

To what extent, if any, this casual recycling within the rain/river/reservoir part of the greater cycle may eventually be harmfully incestuous is not intensively studied although there have been, in rather unusual circumstances, prolonged tests of a scientific nature which do not seem to have produced evidence of ill-effects when coupled with the ingestion of meals prepared from freeze-dried foods.

Cosmonauts, locked in a circuit where economy of the carefully selected scant resources of a tiny environment determine the limits of consumption, pollution and recycling, have thrived as well as survived for some two-thirds of a year, and that on a diet which depended on the same style, but especially more intense, of waste water treatment. Daily requirements of energy and water together, weightless only in orbit, would be quite a 'pay'-load, even if shuttled in several supply missions. So it is reasonable to assume, although not publicly documented, that some means of recycling body fluids through an artificial purifying mechanism (such as a semi-permeable membrane) would be necessary.

Hydrox (Bacon) fuel cells with an energy density some eight times better than that of the more usual silver–zinc cells were used reliably in the Apollo programme. Their by-product, water, *was* used to rehydrate some of that Space crew's rations. The life of a fuel cell is not yet of such durability that it might with surety last for the year or so that Space travel, as distinct from Earth/Moon gravitational journeys, is envisaged to require.

It has been reported (Anon., 1986, p. 42) that 'approximately half of the water demand is provided by the water regeneration system which collects the condensate from the atmosphere in the (space) station; the other half is delivered by cargo ships from Earth'.

As one might conjecture, some of this delivered water could be utilised to back-wash a reverse osmosis device which had become laden with deposits from urinary tracts. Such plants are well known in medical circles as 'kidney' machines, and provide a thorough and effective cleansing action when applied to waste waters external as well as internal to the body. They require lesser back-wash liquid than is passed in the operative direction, so one may infer that for cosmonauts, who have to keep extraordinarily fit in body as well as mind, recycling of body fluids is not stretched to the limit so far. For longer periods further away, space station personnel may well require further refinement, but so far the record admirably indicates that people can live substantially on freeze-dried foods for considerable periods under difficult arduous conditions without any apparent adverse side-effects. Admittedly the cosmonauts are not polluting their water supply with industrial chemicals from engineering or agricultural sources, or neutralising them by addition of yet further substances. All of which leads one to believe that, if there is a fault with rehydration, one must look first to the water rather than the preserved foodstuff.

An average adult human male weighing 70 kg, of which nearly two-thirds is watery liquid, can lose—in perspiration, exhalation and urine—per day at least 1 kg (Hawthorn, J., 1981), perhaps 4 kg if strenuous work is prolonged. Then his regular water replacement needs must be satisfied at least twice a day if severe discomfort is not to be experienced. Heavy work may demand energy expenditure up to 40 kJ/min, so that nutritive assimilation rate determines not only body-building repairs but for how long a determined person can sustain effort. Man will readily consume more than 1·5 kg of balanced food per day if given the chance thus to 'relax', of which 1 kg might well be moisture in integral content. Although the combustible external supply of protein and carbohydrate may thus be provided, another 2 (even 3) kg of liquid may still be required. Such refreshing intakes are taxing, but not considered unpleasant. Nevertheless a stringent dry diet washed down with copious amounts of beverage is reminiscent of horse-power and little-but-often is more in keeping.

Nowhere is this regime of work, rest and diet more clearly demonstrated than in a space vehicle. Limited room for movement is linked scientifically to intense mental alertness. Weightlessness has been found to put the heart and other organs 'out of training', unused muscles weaken, bones need not be so strong. Calcium can then be lost, making return to earthly conditions a hazard. These developments are avoided on long orbital sequences by special suits (Penguin, Chibis) (Anon., 1956, p. 44) to load the body, enabling the cosmonauts to exercise vigorously, for example on a bicycle ergometer.

Soviet space expeditions are getting longer, 140–237 days. At least one crew member, by two expeditions, has spent nearly a year in space, without adverse effect on his health. An interesting, varied, appetising diet is obviously important in maintaining welfare.

Cargo ships shuttle test materials and equipment to and from orbit, and can therefore be called on to replenish food stocks, but water, in or supplementary to food, is an expensive and considerable non-payload. To reduce this burden, a fluid cycle is arranged within the spacecraft, a miniature system of what happens in mundane existence; expelled body fluids are captured, cleaned up, and the water content recirculated. The fresh water brought by the shuttle service is used to back-flush the devices and maintain performance.

Exhaled moisture can be collected by the air-conditioning system, amounting to about half a litre per person per day. Sweat gland secretions, mainly half a litre of water, sodium salt and some urea, are more difficult (cosmonauts shower with soap and sponge in a plastic booth, wearing a face mask). If perspiration is heavy, the pituitary gland warns the kidneys to conserve water and concentrate the urine. However, if liberal drinking were to be undertaken, the urine would become dilute and increase above $1\frac{1}{2}$ litres per day. Selected mineral salts from broken down discarded tissue are also discharged in this way. The hypothalamus balances much more than the body's own water content.

For the menu, purées are just as easy to manoeuvre as pure liquids. All forms of reconstituted food and beverage are available, bearing in mind the complete cycle and admissible occasional excess dietary intake.

The water regeneration complex within the space-ship is presented with a conglomeration of radicals and not a few large molecules. Its task is not just a simple filtration. Moreover, if the tendency is to excrete calcium which should be retained pending the return to gravitational forces, then possibly some hardness in the water of reconstitution is perfectly permissible.

This fits in, in more than one way. People normally accustomed to soft water appear more prone to heart disease than if from a hard-water area. Magnesium and calcium may offer some protection (on Earth of course they can also thwart pick-up of lead and cadmium). If the cell membrane walls within the dried foodstuff to be reconstituted still retain their discernment, a weak solution of metal salts may aid the ingress of the water molecules to the dehydrated interior.

Anabolism is the name given to the general process of building up, and catabolism to breaking down, which continually go on in living cells and composite organisms. The ability to metabolise is not impoverished even when matter as complicated as sperm, for example, is subject to a cryogen, stored for a while and carefully revived. The dormancy is one of degree, and perhaps also of time. Since the general 'memory' of what to do is bound into the nuclei of the protoplasm, there must be a distinction between *how* to build up, which specifies living, and *what* to break down, which may speculate dying.

It is possible to remove most of the water content of a cellular structure as vapour and still allow reversible changes to occur in the protein-containing colloidal parts. The water most easily separated as liquid may already be distinguished by vacuoles, and very fast freezing may lock this as ice within the cells while also imprisoning salts and sugars. If freezing is less fast, then arrangement of the contents can be different, the colloids tending to become more 'sticky' towards the larger molecules, while allowing the moisture to crystallise out where there is room. In Nature this reaction is safeguarding.

If the expression is such that the cell wall fibrils, which in acceptable vegetable foodstuffs are likely to be more pectic than ligneous, cannot bear the transformation, then some irreversible damage will occur. Mostly, however, there is ample intercellular air space to which the moisture can regress, as happens when scalding or blanching upsets the waxy linings.

With raw meat muscle cells to be freeze-dried, a common practice is to chill a prepared carcass first, then butcher it into defined fairly large pieces of tissue, before freezing. Thus the cut meat has both extracellular and intracellular free fluid. The fat vacuoles and the proteins tend to keep the cell gel away from the cell wall, so subliming ice depends on cut cells for main vapour escape paths. The striature is virtually random, unlike in fish, and no long cells survive intact.

Several salts congregate between and beyond cell walls. When drying occurs using end-of-run temperatures which are higher than that of normal bodies, those ions which are left near fat may hasten oxidation although, when a weak solution of such salts is used in reconstitution, rancidity is otherwise retarded.

It is reasonable to aver that the hardness at least of the water of reconstitution is significant.

The cleansing of water in the grand ocean–cloud–rain–river sequence is only truly effective if the air as well as the gravity water system is itself free from defilement. Of course this is not so. The all-pervading by-products of locomotory combustion are oxidised and acidified in the atmosphere, later to leach metals from the soil, nitrates from farms, phosphates from detergents. Eutrophication may please intermediate forms of lower life, but overcoming such pollution by dosing with further, neutralising (in the hydrogen ion sense) pollutants may simply increase a feature of under-proclaimed significance, the insoluble hardness.

Many discharges are legal, including human sewage. Per person a flushing toilet contaminates some 59 m^3 of fresh water a year to wash 3/4 m^3 of sewage further down the line.

The functional feature of communal treatment plants is bacterial action. For this to be effective, limits must be placed on pollutants likely to damage the captive *microbial* population. Some solvents (e.g. trichloroethylene) suspected as possibly carcinogenic, defeat natural sand, soil and clay filters. Waste sludge from effluent treatment is a disposal rather than a recovery problem. Piped water is respected most in areas that do not have it, or rural environments, and water authorities tend to rely on the imperious but evasive description 'potable'! Control, relatively new in consensus, is by maxima toleration, partly because of expense, partly because statistical longevity is a silencing argument.

All this is merely to stress the interesting unconsidered contrast between carefully segregating H_2O as 'pure' ice within a foodstuff to be dried, extracting it with perhaps a few volatiles, as vapour, and then, much later, attempting to (indeed succeeding in) reconstruct(ion) of the stored foodstuff to its original condition. This is done, in fact, by adding a fluid, in the intermediate phase, nominally and principally of identical chemical formula, which—for the most part avoidably—contains also traces of many other substances previously unassociated, in the mind at least, with the foodstuff in question. It is tempting to dismiss this obvious fact as unimportant. Much more liquid, water, is imbibed as beverage regularly than is likely to be taken as part of a reconstituted dehydrated food, and any defects, temporary discomforts or irrational dislikes are minute. The body, with largesse, sends most contaminants on their way. What is of value in the present discussion is whether any special action is significant when 'potable' water meets the first instance, bare, dried tissue.

At least 30% of Britain's population live in areas where the tap water is extremely hard. Government guidance is such that no attempt is made to soften the water before distribution. Indeed, water authorities advise that, if individual domestic purpose supplies are to be softened 'in house', the drinking water itself should not be treated. Renal failure is less prevalent than heart disease! Modern kettles economise in electricity by shutting themselves off whenever the water reaches boiling point; this is not long enough to remove temporary hardness.

Birmingham, with over a million people, gets its (very soft) water direct from mountainous Wales, and discharges, after thorough treatment, into the Tame/Trent rivers, which are not used particularly as water sources, despite passing Burton. These act as refined effluent carriers to the North Sea via the Humber basin. The 'beer' towns draw their supplies chiefly from boreholes. Where 'soft drinks' and beers are produced, individual factories go to considerable lengths to prepare their principal raw ingredient, up to and including charcoal filtration. Contrast this with what is expected of powdered desserts, dried soups, gravy browning, pizza mixes, not to mention the delicate teas, all of which must still taste good, regardless of the local water.

Water *hardness* as a term stems from the tendency of 'hard' water to precipitate the anions of natural soaps as calcium or magnesium salts, thus destroying their detergent or 'wiping' properties. Powder and liquid substitutes are designed to minimise persistent foaming, dissolve fats and grease, and soften water, and so are liberally employed. Exchange softening, using sodium aluminium silicates or various types of resin, replaces calcium and magnesium ions only locally, and moreover only temporarily, since the brine or other wash-back facility re-delivers the (permanent) hardness downstream. Most water authorities find it embarrassingly expensive to soften water at treatment

TABLE 17
Composition of Water Supply

Council of the European Communities Directive (1980), defining the physical, chemical and biological parameters of water quality for human consumption, laid down (amongst other limits):

	Normal restriction per litre	*Maximum allowable per litre*
Aluminium	0·05 mg	0·2 mg
Iron	50 μg	200 μg
Manganese	20 μg	50 μg
Nitrates	25 mg	50 mg
Ammonia, NH_4	0·05 mg	0·5 mg

	Maximum concn. per litre
Lead	50 μg
Mineral oils	10 μg
Nitrites	0·1 mg
Surfactants	200 μg

Suspended solids: none
Pathogenic organisms: none

Insecticides, herbicides, fungicides and the like: individually below 0·1 μg/l; in total below 0·5 μg/l

pH between 6·5 and 8·5

works; they have plenty to do without that feature but they do adjust pH, settle out bicarbonates by adding alum early, and adding lime later on. Hard waters tend to prevail (Bohn *et al.*, 1981). The pH-buffering capacity of natural waters is almost nil.

The Severn–Trent Water Authority tends to the needs and disposal misdeeds of over 8 million people. Even in a comparatively well financed inland area of the UK, the achievement of EEC standards for typical potable water supplies has been uphill and thought-provoking (Table 17).

Summarising, it has to be conceded that, instead of the 'H_2O' which was in isolation extracted, the liquid of reconstitution has to be that which is conveniently available for domestic or at least culinary use, a potable complication of dilute solutions, the technological result of intensive urban spread. So habit and necessity have inured us to the daily imbibition of (albeit partially) degraded waters, which are assuredly safe and free from bioaccumulative toxins, but perforce vary clearly in faucet districts, in content, colour and taste.

Since opportunity to compare organoleptic experiences tends to be confined to immediate neighbourhoods and small-talk, and even then centres on the flavour of beverages rather than 'solid' foods, no great reaction, favourable or otherwise, need be expected. Water is taken for granted as an ingredient of communal society.

However, if some catastrophe upset the network, such as a nuclear burst which would imperil all kinds of logistical supplies, then deep borehole water and concentrated stores of dried foods might be combined for the common good and survival (Chapter 12).

Chapter 12

Why Ever Not?

ATMOSPHERIC FREEZE-DRYING

Freeze-drying at atmospheric pressure commonly refers to the condition whereby the combined pressure (and presence) of non-condensable (not necessarily inert) gases is some 200 times greater than that partial but additional pressure of the water vapour over the subliming ice. While this presence does not 'interfere'with the phenomenon of drying, it does have an associated effect.

To keep the numbers simple and familiar to processors, mm Hg will be used in this context. Thus a normal atmosphere at 0°C is 760 mm Hg = 101·325 kPa, while vapour pressure is as indicated in Table 18. While barometric pressure may vary between (say) 760 + 30 and 760 − 50, partial pressure of water vapour can only vary by (say) 32 mm.

The popular illustration of atmospheric freeze-drying depicts wet clothes or blankets hung out on a washing line on a very cold bright day. The water soaking the material evaporative-freezes and then sublimes, while the stiffened cloth gradually dries out and becomes flexible again. The special conditions are inferred to be (a) thin matted-fibre material of large surface area, (b) exposure to really cold wind, i.e. moving air in the open, (c) radiant heat from an unobscured sun or from black-body mother earth.

While these conditions lead to rapid results, they do not stipulate the major aids to the phenomenon, which are, in simple terms, that (a) the temperature immediately *surrounding* the 'clothes' should remain below 0°C, and (b) the subliming vapour should have a clear sky to go to, i.e. a 'sink' where the partial pressure of water vapour is

TABLE 18
Variation of Partial Pressure of Atmospheric Water Vapour

Temperature (°*C*)	*Pressure of water vapour over water*	
	(*mm Hg*)	(*Pa*) (not *kPa*)
+30	31·8	4239
+20	17·5	2333
+10	9·2	1226
+5	6·5	866·5
0 over ice	4·58 over ice	610·5 over ice
−3	4·0	533
−5	3·0	400
−10	1·97	263
−15	1·3	173
−20	0·79	105
−30	0·27	36·5
−39	0·1	13·3

even less. This is not difficult; moving air will tend to be of lower total pressure, allowing water vapour to expand itself to a lower partial pressure equivalent to that saturating a space at a correspondingly lower (ice) temperature or a relative humidity *at* the actual air temperature.

Continuing the example, at −3°C a keen housewife might brave it outside in the UK, but might have doubts at −20°C. In practice, the extreme partial pressures of water vapour surrounding and in the material being dried will lie between 4·58 mm and 0·1 mm Hg for freeze-drying to take place, and towards the latter value for it to continue until the deepmost ice has sublimed away, because of interstice resistance. The surrounding air has of course to be receptive to the emerging vapour. For ice at −3°C (say) and an air temperature of −5°C but also a relative humidity of (say) 45%, reference to Table 18 gives the partial pressure of water vapour in that air as close to 1·3 mm, so that not only can it accept and contain more vapour but there is a pressure gradient from the ice to the outside 'air' inducing the vapour to proceed there. The rate at which this will happen will depend on the heat of sublimation getting to the ice.

Although the *level* of the heat is low in general experience terms, it is not low in absolute terms, and the amount of heat carried by the surrounding air is substantial compared to the finite amount of heat embodied in, and to be extracted from, the material. It is to be remembered that, although no heat will flow between substances axiomatically at the *same* temperature, only a very tiny difference in temperature level is necessary for it to flow by convection or conduction. Obviously an external radiant source will also lend a hand. Such a difference occurs in all drying procedures, instituted in this case by the partial pressure difference demanding molecules from the ice, causing its internal heat content to fall, hence its temperature, hence its ability to release more molecules. In so doing, the partial pressure of water vapour immediately over the ice will fall, until there is no vapour pressure difference left, at which point the drying process will cease, that is unless more heat arrives to replace that which was lost at the ice front, maintaining the temperatures there and therefore the pressure gradient.

Such heat will be obtained by all three forms of transmission, through the air both outside and inside the fabric, but most assuredly by conduction through the body of the fabric. The respective rates at which these modes operate do not much matter if the clothes dry out in the reasonable time of a few hours, or more likely in less time; no one is likely to ponder over it in a busy household, nor is the fabric likely to suffer any physical duress from the experience.

It is also a matter of common experience, and therefore a suitable reference, for foodstuffs to be noticed as having dried out or lost weight in the domestic refrigerator. Opening the door frequently lets the cold air fall out and warmer room air sweep in, and the ice accumulating on the freezer plate or coils is mostly from moisture extracted from this room air. But some of the ice will have come from moisture in the foodstuffs, migrating even when the door is not opened for long periods. The temperature in the main part of the refrigerator will probably be around +4°C and the foodstuffs not frozen, but the migration would occur if they were. In a deep-freeze conditions can occur, or be arranged, so that sublimation does take place, very inexpensively.

In extremes of climate not ordinarily experienced other than by mountaineers and explorers forgetful of the danger, freeze-drying of parts of the body exposed inadvertently can occur. Frostbite can be alleviated by rubbing the affected part *gently* with snow at the same temperature, not to supply heat as much as to replace the lost moisture. If the frostbite is severe enough that blood supply is completely lost, mortification can set in, but within limits the process is reversible and the tissue restorable. It does not take long for frostbite to set in; presumably the heat of the body itself feeds the process, the bitter cold of the air, from which all moisture has been extracted as snow, representing a very low humidity and therefore low vapour pressure; the 'sink' is deep, the action unpleasantly rapid. Not all creatures are as fortunate as Merryman's mouse! In this classic experiment into treatment for frostbite, an American researcher successfully freeze-dried then reconstituted the still-attached tail of a mouse with no apparent harmful after-effects.

It will be clear from the foregoing that daylight as such is not necessary, although radiant heat from a wintry sun would speed some of the processes outlined. Despite the vagaries of the weather, it is feasible to view atmospheric freeze-drying from a practical commercial angle, provided one can afford to wait while relatively minute forces are acting, and at no time allow the ice temperature to rise such that the partial pressure of water vapour above it exceeds 4·58 mm Hg. With Nature providing the outlay, the freeze-drying process will happen anyway, as herring hakes hung in winter on the gables of fishing village houses used to testify.

Rarely can one envisage a commercial process wherein time is not a pertinent factor overriding most others. An interesting parallel is that 'National Food Situation' published by USDA Economic Research Service 1977 cites Bill HR 6014, introduced by Representative Beard of RI: 'To establish an emergency human nutrition food reserve which shall be able to help meet emergency food conditions in the United States *or any foreign country* which suffer a severe loss of its food supply as a result of a national disaster'. The most common criterion is running cost versus increased capital cost; in times when capital is expensive, simple systems which take longer are sometimes retained beyond the point when further or newer mechanisation would seem to be dictated. Terminal worth is never easy to measure, but the cost of the risk of retention presents less fear than the cost of capital for the new. Fortunately this broadly has not tended to be the general case during enthusiastic periods of scale-up involving freeze-drying of foodstuffs.

There is, however, a situation that is ready-made for atmospheric freeze-drying, namely cold-store operation. There used to be reported in cold-stores an alarming phenomenon termed 'freezer burn', whereby surface desiccation discoloured and dried out cuts of foodstuffs exposed to the dryness of high-speed cold-blast air. Lamps which could radiate heat through this cold-blast air, without warming until the energy reached solid objects such as frozen foods, and then really did not warm but fed in the heat of change of state, were part of the unwitting aids to the sublimation. A certain amount of 'toughening' was reported if the process was allowed to continue, probably above ice-melt temperatures, but this effect was more severe if air speed was too rapid rather than too prolonged.

In general, particularly because of loss in weight of product, which previous added water did not quite alleviate, 'freezer burn' was not acceptable. The method of avoiding it was to package the material tightly in moisture/vapour-proof wrap, thus inserting a barrier in the partial-pressure gradient between the water vapour above the ice-front in the foodstuff and that above the ice on the refrigerated coils of the evaporator in the blast-air system.

Possibly these avoidance tactics would have been the end of the matter had it not been for the curious outcomes of the EEC Common Agricultural Policy. The desire to sustain certain agricultural activities regardless of open market need had resulted in surpluses being hoarded in cold stores all over Europe for long periods.

The cost, communally borne, of such storage alone is mountainous, and means of reducing it without necessarily otherwise altering the policy has been sought. If the produce could be kept in ordinary temperature stores, this presumably would be less costly than in the special, insulated, engineered-environment stores, and if the produce was *freeze-dried* most of its life in limbo could be spent in the simpler store. Further benefits appear as soon as the courage to consider the proposition as realistic has been summoned up. Beef may be considered as a prime example.

The EEC guarantee farm prices, buying up agricultural and dairy produce if the market value does not reach an agreed figure. Beef cattle, produced beyond the public demand quantity, are purchased, slaughtered, butchered, boxed in selected cuts, frozen and stored at −30°C, all by the communal purse.

There it will tend to remain for long periods, upwards of 10 months, at a cost (1986) in the region of at least £1 per tonne, *per week*. In accordance with contrived rules, it must be disposed of outside the EEC, which means that it must compete in external, world-wide markets, or be sold off at less than its value-added cost. Almost invariably the inherent cost is greater than its worth in *boxed-beef* form, and no profit is achievable to the taxpayer in the end transaction.

Periodically its presence becomes an embarrassment and it must be moved. Some cold-stores, nominally public stores, are 70% full of 'intervention' foodstuffs, which may be disposed of abruptly, leaving a gap in the cold-store rented holding which may be difficult suddenly to fill, and such risks no doubt must be allowed for in the rental charges.

To a certain extent freeze-drying can help in this situation. The beef cuts may be thought of as 'retail' cuts and 'commercial' cuts, the latter, for example, including joints and other sections intended for soups or pie fillings or comminution. If such meats were already in minced or granulated form, cooked and freeze-dried, then they would command, in a relatively narrow but surprisingly extensive market, a much higher price than as commercial joints. Such further cost, on top of an already artificially high raw-material cost, would make the disposal price even less commercially acceptable, that is unless the expense of dehydrating was contained in a sufficiently shorter period in the cold-store, and the commercial cuts were then more easily stripped of meat and dried on the bone ready for dry mincing. (The retail cuts can of course be dried as such, for eventual reconstitution and marketing as dried, a separate scheme; their sojourn in the cold-store need not necessarily be markedly different from the commercial cuts.)

A few general points on departure from previous practice are as follows:

(1) The cuts should be roughly identifiable in drying thicknesses, not as a complication, but to make it easier to establish which are ready for the next stage. Time in store to dryness can be as short as one-tenth of that previously employed.
(2) During freezing and cold-storage, the cuts should not be wrapped. Toast-rack formation would obviously be beneficial, and make for uniform manipulation.
(3) Insulation of the shell of the store by hermetic refrigerated jacket not only slows heat ingress, but evens it out. Stores vary with age in heat-tightness. The interior of the shell should be looked on to provide useful low-temperature radiation.
(4) Under-floor 'anti-frost-heave' facilities should be brought into play as the main radiant heat source. Direction of air currents should be kept in mind.
(5) The 'ice condenser' is the evaporator coil(s) of the refrigerating machine. Build-up of ice will affect the partial-pressure gradient, and must be minimised, so frequent defrosting *which does not upset store conditions* must be provided. Systems that reduce

build-up by intense air-blast, for instance, must also be isolatable. The ice should *depart* the system.

(6) The produce must be overwrapped before being removed from the dry atmosphere of the 'cold store', now acting as a process area, at the same time being checked for dryness using modern rapid techniques. Thereafter it can be graded and further processed as required. When in its final vapour-proof wrap, it needs only ordinary temperate storage facilities.

(7) The cold-store is preferably leak-tight. While oxygen content of the store atmosphere may be necessary for operators' convenience, depending on the 'drill', it is not of particular value in a meat store.

(8) Although beef has been put forward as an illustration, other produce can be treated similarly.

Purging atmospheric O_2 down to trace conditions, and holding, is probably easier and cheaper to achieve, in a tight store, with CO_2 rather than N_2.

In 'energy conservation thermographic surveys' aerial infrared photography is used to delineate areas of (industrial) buildings that have little or no thermal insulation, and therefore lose costly heat to the atmosphere wastefully. Shades of grey distinguish relative rates of heat loss by representing (roof) outer surface temperatures. Furnace rooms and steam pipes show up distinctly but, even where only comfort heating is in vogue, some gradation can be seen on the imagery.

Unfortunately the technique is not employed in reverse on cold rooms or refrigerated stores—not by aerial viewing but from inside—as many early editions of these are still not too well clad. Wherever a temperature gradient exists there will be a flow of heat; the amount and nature of insulant merely adjusts the rate. Single-shift occupational requirement, with elevated temperature for only 25% of the week, is a different problem from the same difference (albeit at lower level), all 168 hours of it. Seasonal effects are opposed when the unwanted 'free' energy comes from outside; diurnal variation may aid but, if the store is to be involved in the freeze-drying operation, insulation of the most modern energy-restricting design is advisable.

A temperature difference of 30°, for example, given an external average winter temperature of (say) +7°C, is large for normal workplace buildings, but small for a cold-store. Therefore one must consider U values below 0·25 W/m^{2}K. Even at that level, and assuming a shape 100 m square by 10 m high, 185 kW is not (in winter) an exceptional heat intake, or the 65 kW plant call unusual, which deals with it by air-circuitry, maintaining the temperature of the goods inside at the example −23°C, jacketed or not.

If it were arranged and allowed to, such an evenly distributed amount of energy could sublime ice at −23°C at a rate that would dehydrate the store contents comfortably in weeks rather than months. It is a question of attitude rather than engineering. The Egyptian pyramid/tomb builders were thoroughly conversant with (a) insulants such as sandstone and flaxen bandages, (b) water activity effects of embalming fluids,† (c) really dry still air, but then they *wanted* to preserve in that fashion. In modern times, where the cold chain predominates, it is not conceived as an obvious lateral-thinking step that drying-out should follow segregation! Cryptically, the re-reader may rail against Smithfield cheesecloth in favour of (plastic) expanet sleeving!

Summarising, it is possible to utilise existing cold-stores for atmospheric freeze-drying, so that they become, with minor adaptation, intermediate process plants, having a turnover of produce up to ten times that of 'normal' intervention holding, with subsequent storage in less expensive conditions, and quite possibly up-grading the produce in so doing.

THE WHOLE HOG!

It has been reported (Geraghty, T., 1983) that, during the 1982 Falklands campaign, 'each (SAS) team abseiled (from helicopters) into squelching bog . . . to be staked out, immobile, in freezing peat. A fourteen-day supply of rations, a meagre dehydrated diet, had to last 26 days. The trench-foot most of them contracted was reminiscent of Flanders'.

This gangrenous affliction, brought on by prolonged exposure to damp and cold, causing excruciating, crippling pain, and not unlike frostbite, was a major theme at a conference of the Royal Society of Medicine on the lessons of the Falklands campaign. It was noted that

† Examination of the collagen tanned remains of Lindow Man ('the body in the bog') by the British Museum (Omar & McCord, 1986) involved pretreatment in a 15% solution of polyethylene glycol 400 for 10 weeks, freezing to −26°C for 3 days, freeze-drying at −20°C (condenser −40°C, cabinet 0·15 mm Hg) over 4 weeks, then equilibrating to 55% RH at room temperature.

the 'boot, ankle, direct moulded sole, general service', used by the British Forces at that time, let in water at the lace openings, had low insole absorbency, so would not dry out nor 'breathe'. Some fortunate troops had the higher, more waterproof, Cairngorm or Greenlander boot, some wore Swedish boots, some plastic overboots, some even 'borrowed' Argentinian issue boots, all of which were much more expensively produced, but gave no such trouble.

Hot-air blowers from field hospitals were used to dry out *feet* and restore circulation, but rather too many personnel had to withdraw from active operation, partly because of this and partly from dysentery through melting dirty snow for use with their dry-ration packs.

Major-General R. Wall, Royal Marines, reported that specification of a new 'boot, combat, high leg, DMS' should be more suitable for all sorts of climatic conditions, but General Sir Jeremy Moore, Falklands commander, reckoned it was *not* possible to get the *perfect* boot until one could breed the perfect cow!

Without resorting to biogenetic engineering, as long ago as the 1950s whole cow-hides were successfully freeze-dried between 8 ft by 4 ft double-contact AFD heating plates in Aberdeen. The result was a breathing, flexible skin whose collagen was capable of taking an impregnation of (vacuum-vaporised) resinous type coatings. Unfortunately this equipment, still in its experimental state, was not adaptable to follow through with the idea of non-food use, but the writer, 30 years on, has still in his possession, to ponder over, in perfect condition after virtually a generation in the ordinary air, a small sample of what could be produced.

THE OVERVIEW

World food supply information is collated for the Statistical Yearbook of the United Nations Organisation (UNO, 1984). Such an immense preliminary task of engendered co-operation depends on summations whose value rests on some regularity in the level of declared accuracy, so that general trend(s) may be capable of dependable inference and the comfort of charitable action taken.

Thanks to the Food and Agricultural Organisation of the UNO, the famine and undernourished areas of the world are well known and monitored, as are the regions of glut. The foodstuffs in surplus tend to be cereal grains which are not habitually used in the famine relief

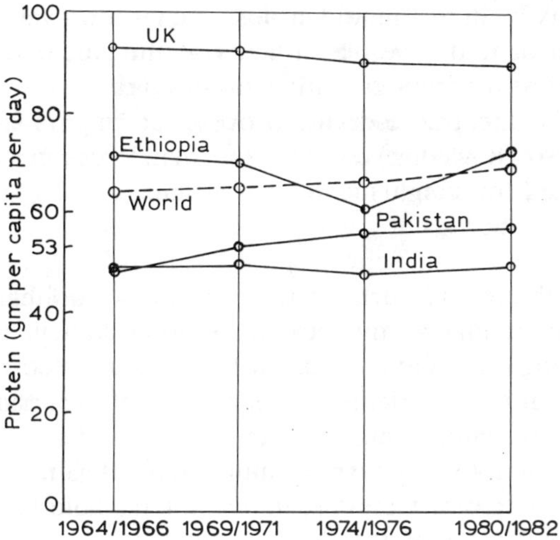

FIG. 61. The protein intake imbalance (e.g. India has a population 12 times that of the UK).

areas except by gift. Shortage of protein is *cyclically* related to shortage of water. *Gruel* becomes the *dependency* diet!

In Britain (population 56 million), the mean recommended consumption per (male) person per day is 2160 cal, 53 g of protein, and we find (Fig. 61) that the averages *per capita* reported are in the main well above that. In fact we have to note that India, with a population of some 684 million, is reported nearest to this sample norm, although steadily just below it. While the world trend is commendably rising, the UK trend is a slow fall; both are well above the UK dietary target(s). A healthy outlook?

Charter It!

Transcontinental cargo flights are no longer rare. Economically, trips depend not only on payload but on height. At an altitude of say 12 km, threshold of the stratosphere, the external temperature will be lower than −50°C, equivalent to a partial (water vapour) pressure of about 0·03 mm Hg. Not only will foodstuffs such as mango pulp, passion fruit juice, and so on, in a non-pressurised cargo-hold freeze but, given merely cheap waste heat from the aircraft, suitable sectioning and packaging, it will *freeze-dry,* and that quite rapidly. Some journeys between refuelling stops are still of ample duration for such a facility, where the weight-loss curve soon tempers the initial (take-off) burden, or even an otherwise unavoidable empty return-flight hold. Whilst wintry conditions at ground level can easily provide the format for the freeze-drying process, at heights less than that of Mount Everest (8·848 km) *anywhere* a partial pressure differential can be experienced (or sought) which will suffice.

Barter It?

In parts of the world such as the Ongallala aquifer for the North American Great Plains, and the San Joaquin valley, where intensive irrigation from underground reserves has so affected the water table that substantial land settlement is in danger of occurring, the practice of transporting water supplies from where it gathers to where man gathers is presenting long-term complications. Again, the international Libyan project, piping desert region aquifers, holding water equal to 200 years total flow of the Nile, to irrigate some 1·2 million km^2 around the Gulf of Sidra, may bring back Carthage, but increase troubles further south.

Capture It?

Pollutants in drinking water are sometimes significant even when only mere fractions of a part per million, and can only be analysed, identified and measured after concentrating the sample in an unaffecting way, i.e. by *freeze-drying*!

The FAO returns are based on amounts of foodstuffs not just 'available' but actually 'consumed', e.g. homeland production plus imports less exports, less storage and distribution losses, less processing waste, retail trimming, deterioration, and less that fed to pets, leftover scraps from the plate and so on! Then a calculation, weighted for infants, youngsters, able-bodied, elderly, ill etc. provides a notional population. The quotient, translated into nutritional units, gives an overview of the possibility of comparisons such as Fig. 62.

On the face of it (Fig. 62) there is no cause for alarm, despondency or additional action. The basic freedoms appear satisfied. After all, human nutrition overall depends on little more than 30 plant types, and 8 of these in present production could provide three-quarters of the arbitrary calories.

However, imbalances prevail which 'totals' and 'averages' do not

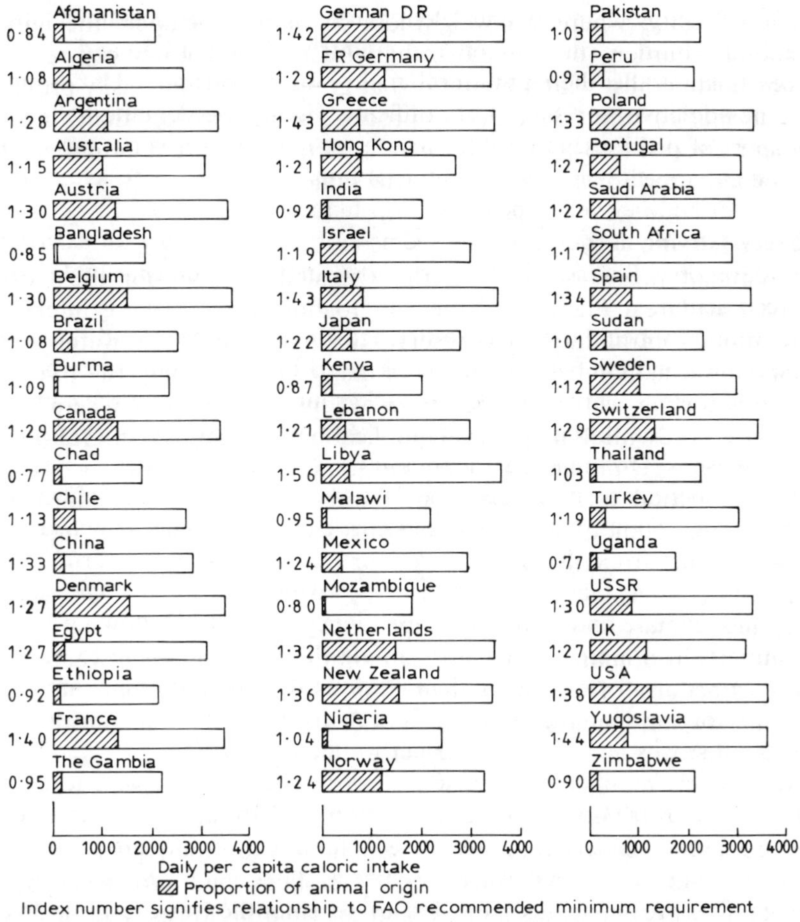

FIG. 62. Comparison of diets by country, *c.* 1981. (Data used in this figure has been adapted from the *Encyclopaedia Britannica Yearbook 1987*.)

alone reveal. Traditional methods of drying and storage, where these apply, may be labour-intensive but preferable to dependency on borrowed capital or higher technology. Some countries are known to export vegetables, for instance, while importing insufficient grain. Rice may be stockpiled when undernourished and flood-stricken locals cannot then afford it. A persistent complaint is that over half the cultivable land in the Southern Hemisphere resists reform to be available as small-holdings inferentially eager for higher yields of more 'at home' crops. Famine and plenty are reported as sustained and tolerated within a single wretched country when market and other forces 'justify' the then unnatural nature of catastrophe. Hand-outs become addictive, self-help more difficult, dignity too demanding. Fast transport of precious perishables to a beneficial market is an aim with immediate appeal compared to internal preservation so as to avoid the ravages of another year's possible crop failure.

Freeze-drying in such contexts seems morally viable only for *external* consideration. Statistics duly precede inexpensive charitable or coerced action.

It would appear that freeze-drying, or any other food quality preservation system which maintains nutritional content as a paramount feature, is only coincidental background colouring to the canvas that receives the broad brush of the FAO/WHO strategies by which the UN seeks to make a clean sweep of the anomalies it faces. Over-production of basic produce is a delicate subject on which to warrant suggestion of additional and particular refinement expenses.

Quite apart from the evident measure of success at the farm gate (on which many factions lean), the various needs of age, sex, physique, climate and economic dependency are in the follow-up level of interest. The mere fact of altruism via the UN being able to cajole such figures at all from other than obvious donors and recipients is a triumph for the simplification that had to ensue in order to start it off. Foodstuff stocks however are subject to deterioration.

Within this ramified sphere of UN work is a laudable aim to seek more even distribution (sometimes simply distribution) of declared and apparent surpluses. Cereal seeds tend to be the international currency, and nutritional variety is hence its problem. Seasonal glut and famine are tenaciously located to be smoothed away, while loss and waste are admittedly otherwise enormous. Such preservation processes as freeze-drying do not fit the major items. Also 'food' supply has become a munition of bargaining power, and from a business viewpoint ecology—the study of organisms in relation to environment—does not entertain the concept of ethnic Man as an endangered species, especially on averaged information.

An interesting side-light on this is shown by an operational research paper (Anderson & Earle, 1983) on diet planning in the Third World, wherein 'linear programming' of the formulation for a (relief) food product, towards a 'least cost mix', is contrasted with 'goal programming' which optimised the nutritional balance, albeit at some extra expense. An acceptable combination of the two factors is called for. The pattern of amino acids in the (freeze-dried) hen's egg, close to human dietary protein requirements, might thus enter an argument otherwise denied.

The FAO computerised supply/utilisation accounts (SUAs) cover 300 primary food, agricultural and fishery commodities, 310 derived processed products, for 200 countries, over the last 25 years. Supply of foodstuffs is reckoned from production plus imports. Utilisation deducts exports, livestock feed, seed use, industrial and other non-food use, losses during storage and transport. Food supplies available (*sic*) at retail or entry to households are divided by related population data to give *per capita* food-item composition in terms of nutritional elements. Losses during storage thereafter, preparation or plate-waste etc., are not indicated, nor are differences in diet of different population groups and zones within a country. Seasonal variation has to be inferred. One is tempted to conclude that a major benefit sought must be realisation of how brave such fragile statistical approaches are, and just how vulnerable mankind still is to the vagaries of political climate!

The Final Say!

At the 1984 annual meeting of the British Association for the Advancement of Science (*nullus in verba*) a spokesman from the CEGB forecast that an extra 9 gigawatts of generating capacity, on top of the then current 52 GW, would be needed by the end of the century, allowing for obsolescence of existing plant. Apart from dependency on coal versus nuclear versus oil fuelling considerations, it was put to him that conservation was more cost-effective than building extra capacity, and moreover that computation of electricity usage by fridges and freezers in the UK, for example, amounted to the output of three Sizewell 'B' reactor power stations (Barrett, M., 1984, pers. comm.).

By inference, increasing the overall efficiency of the nation's fridges could save 2 GW, i.e. an extra power station; so, of course, could increased use of freeze-dried materials, whether the process was carried out in the UK or in a supplying country.

No apology is made in this book for omitting any discussion on 'cost', that four-letter word constantly on the lips of all professional engineers. Despite the cautious attitude of venture capital, headway has been maintained, and the products found secure places in the market perfectly willing to accept the price of quality.

Why then should foodstuffs be preserved by a means that is capable of maintaining for many years a standard of acceptable quality in an otherwise perishable commodity, when presumably dealing abstractly in futures is simpler, and equivalent quality may be achieved continuously without storage by rapid round-the-year supply from successive parts of the globe that are rotationally in harvest season? As far as Man's rationale of commerce allows, this is already done.

People without food at all, of course, rarely have left other resources to bargain their way out of the problem. Discernment of quality of life in others is measured against one's own dispassionate lot; the tallyman takes the tithe because human endeavour is comfortably too involved to be direct. So abundance and want exist economically together and well apart from conflict!

The substantive reasons for freeze-drying foodstuffs remain investment in (a) personal convenience of stored true wealth, (b) rapid realisability of staple necessities, and (c) security from supra-national calamity—and these are enough.

References

Anon. (1981). *Fundamentals Handbook*. Institute of Refrigeration, Chpts 16, 17.

Anon. (1984). Preconcentration for Freezedried Entrees, Paper: presented at meeting of R & D Associates, US Army Natick R & D Command.

Anon. (1985). Thirsty Sailor saved me from freezing to death. *Weekend,* 3 July 1985.

Anon. (1986). *USSR Yearbook*. Novosti Press Agency, Moscow, pp. 42, 44.

Aberdeen Press & Journal (28 March 1960). Your Menus for the Moon, Sir.

Anderson, A. M. & Earle, M. D. (1983). Diet Planning in the Third World by Linear and Goal Programming. *J. of the Operatnl Res. Soc.* **34,** 9–16.

Ackroyd, K. (1988). The Chemical Companies' Plans & Intentions. *Proc. of the NEDC Conference Refrigerants and the Environment.* Institute of Refrigeration, Carshalton.

Barrett, J. R., Laxon, R. & Webster, P. H. N. (1964). Freezedrying Vacuum Equipment. *Food Technology,* January 1964, 38–41.

Bell, G. A. & Mellor, J. D. (1986). Development of the Adsorption Freezedrier. *CSIRO Division of Food Research Quarterly,* **46,** 56–8.

Bohn, H. L. & Cauthorn, R. C. (1981). Pollution: The problem of misplaced waste. In *Use & Misuse of Earth's Surface* ed. B. J. Skinner, Kaufmann, California. Also *American Scientist* **60:561** (1972) 36–40.

Daily Mail (18 July 1956). Dinner is Served.

Decareau, R. V. (1985). *Microwaves in the Food Processing Industry*. Academic Press, New York, pp. 30–7, 124–9.

Evans, H. (1978). *Vickers*: *Against the Odds* (*1956–77*). Hodder & Stoughton, London.

Geraghty, T. (1983). *This is the SAS*. Fontana/Collins, Glasgow, p. 146.

Greaves, R. I. N. (1954). Theoretical aspects of drying by vacuum sublimation. In *Biological Applications of Freezing and Drying,* ed. R. J. Harris. Academic Press, London/New York, p. 87.

Greaves, R. I. N. (1962). Vacuum Spray Freezedrying. *Proceedings of the 5th Freezedrying Conference*. Leybold-Hochvakuum Anlagen, Cologne.

Hanson, S. W. F. (1961). *The Accelerated Freezedrying Method of Food Preservation*. HMSO, London.

Hawthorn, J. (1981). *Foundations of Food Science.* Freeman, Oxford, pp. 2–8.

Inglesias, H. A. & Chirife, J. (1982). *Handbook of Food Isotherms.* Academic Press, London.

Hendel, C. F., Legault, R. R., Talburt, W. F., Burr, H. K. & Wilke, C. R. (1958). Water Vapour Transfer in the In-Can Desiccation of Dehydrated Foods. *Proc. of the Society of Chemical Industry Conference 25/27th March 1958.* Soc. Chem. Ind., London.

Klicka, M. V. (1969). Space food development. *Proc. of symposium on Feeding the Military Man,* ed. M. S. Peterson. National Academy of Sciences, in conjunction with US Army Natick Laboratories.

Labuza, T. P. (1971). Kinetics of Lipid Oxidation in Foods. *CRC Critical Reviews in Food Technology,* pp. 355–401.

Lebedev, V. (1985). Diary of a cosmonaut, *Science and Life,* **12,** 34–44.

Lefever, S. E. H. (1960). Progress Report from Britain to the R & D Associates of FCI meeting, Chicago, 20–21 September 1960.

Lyle, O. (1968). *The Efficient Use of Steam.* HMSO, London, Chpt. 26.

MAFF Press Notice (26 January 1961). Sale to Unilever. Research and Experimental Factory, Aberdeen.

Mellor, J. D. (1978). *Fundamentals of Freezedrying.* Academic Press, London.

Millin, D. J. & Bradford, L. (1984). *Food Industries Manual,* 21st Edition, Leonard Hill, Glasgow, pp 197–207.

Omar, S. & McCord, M. (1986). *Lindow Man.* British Museum, London, pp. 17–20.

Rey, L-R. (1959). Brevets d'invention. Centre National de la Recherche Scientifique, Paris.

Rey, L.-R. (1962). Influence of the preliminary freezing period on Lyophilisation. *Proc. 5th Freezedrying Conf.* Leybold Hochvakuum Anlagen, Koln-Bayental, pp. 3–19.

Robertson, A. (1970). *The Lessons of Failure.* Macdonald, London, pp. 34–46.

Rowland, F. S. & Molina, M. J. (1974). Ozone Depletion Hypothesis. *Nature,* **249,** 8–10.

Scottish Daily Mail (17th July 1956) The Meal in a Matchbox.

Seltzer, E. (1960). Communication to R & D Associates of Food & Container Institute, Internal Report of Meeting 20–21 September, 1960.

Shalliker, J. (1987). Fleet Structures Model. *J. Opl Res. Soc.,* **38** (11) 1007–14.

Sunderland, J. E. (1982). Microwave Freezedrying. *J. of Food Process Engng,* **4,** 195–212.

UNO (1984). World Food Supply. *Statistical Yearbook 1983/84,* Section 95, p. 583.

Walton, A. J. (1983). *Three Phases of Matter.* Clarendon Press, Oxford.

Further Reading

NASA Facts, an educational publication of the National Aeronautics and *Space* Administration NF-150/1–86.

Huttenbach, F. C. (1987). *An Introduction to Food Management in Space.* Paper presented to the Food Science Dept, South Bank Polytechnic, 5 May 1987, Nelson Space Services, London.

And for the Andromeda Types, eager to be spirited away metaphysically *and* metabolically, in a red shift from an immediate local sea-shore plight by the winged-sandalled Aurora, the thought-provoking *Sky at Night* by Patrick Moore (1989), Harrap, London, may well give them food for thought.

Index